NEW PROFITS IN

GOLD
SILVER
STRATEGIC METALS

NEW PROFITS IN
GOLD
SILVER
STRATEGIC METALS

THE COMPLETE INVESTMENT GUIDE
PETER C. CAVELTI

McGraw-Hill Book Company

New York   St. Louis   San Francisco   Bogotá   Guatemala
Hamburg   Lisbon   Madrid   Mexico
Panama   Paris   San Juan   São Paulo   Tokyo

Published in the U.S. by arrangement with
Maximus Press Limited
328 Palmerston Blvd.
Toronto, Ontario
M6G 2N6

First McGraw-Hill edition, 1984

123456789DOCDOC87654

Research assistance provided by the international department
of Guardian Trust Company, Toronto

All information is based on sources considered reliable.
Recommendations contained in this publication reflect the views of the
author and are not to be understood as offers or solicitations of offers to
buy or sell the investments described and are in no way guaranteed.

Cover Design: Douglas Frank
Photography: Davidson-Samson Studios
Index: Christopher Blackburn
Charts: James Loates
Colour separation: Herzig Somerville Ltd.

ISBN 0-07-010288-0

Library of Congress Cataloging in Publication Data

Cavelti, Peter C., 1947–
  New profits in gold, silver, strategic metals.

  Bibliography: p.
  Includes index.
    1. Precious metals—Handbooks, manuals, etc.
  2. Strategic materials—Handbooks, manuals, etc.
  I. Title.
  HG261.C38   1985   332.63   84-12268
  ISBN 0-07-010288-0

To my parents
for always letting me be myself

By the same author

HOW TO INVEST IN GOLD 1979, revised 1981.
Maximus Press Limited, Toronto
Follett Publishing Corporation, Chicago

Socrates and the Economist

Socrates: I see that your chief piece of money carries a legend affirming it is a promise to pay the bearer the sum of one pound. What is this thing, a pound, of which payment is thus promised?

Economist: A pound is the British unit of account.

Socrates: So there is, I suppose, some concrete object which embodies more firmly that abstract unit of account than does this paper promise?

Economist: There is no such object, O Socrates!

Socrates: Indeed? Then what your Bank promises is to give me another promise stamped with a different number in case I should regard the number stamped on the promise as in some way ill-omened?

Economist: It would seem, indeed, to be promising something of that kind.

This imaginary conversation between the philosopher Socrates and an economist appears in "Essays on Monetary Theory", by Prof. D.H. Robertson — published by P.S. King & Son, London, 1940.

Contents

About This Book

The Chinese symbol for crisis consists of two characters, one meaning "danger" and the other "opportunity". We are still in the first half of this decade, but already everything points to the conclusion that the Eighties will be a time of extreme instability. And what could be more rewarding during such a critical period than understanding and exploiting the price trends of those investment vehicles which are the very barometers of economic and political uncertainty: precious and strategic metals.

Hundreds of books and theses, and thousands of newsletters and reports have been written about this subject. Most deal with the chemical properties or applications of a metal, describe its history, or specialize in advocating one investment vehicle over all others. As a professional, I find all these works very interesting, but most people have no practical use for the findings of a chemical engineer or historian; instead, they want to know how to put their money to the best use.

And making the right decision when investing in precious and strategic metals is not an easy task. Professional dealers have responded to the explosion in investor demand and the accompanying unprecedented price volatility by creating a confusing array of new investment vehicles. Besides the traditional alternatives of bullion, storage certificates, mining shares and collectibles, there are now such things as options, metals accounts, toll-free telephone trading facilities, and even savings passbooks denominated in gold!

The purpose of this book is to illustrate the way in which each alternative has been designed to cater to a specific investment objective. It will also detail the advantages and pitfalls to be found in each of the strategies available to today's private investor.

There are also significant differences between the groups of metals we call "precious" and those which are termed "strategic". Both have drastically different characteristics and respond to different events and developments. And even within each group you will find metals which suit your particular investment strategy, and others which don't. This book will familiarize you with the economic and political circumstances which cause each of these metals to decline in value, as well as with the developments which can lead to ongoing price appreciation.

But no book about a financial market would be complete without a close look at those who make it. In the course of our study, you will learn about the feud between fundamentalists who predict market action by trying to understand the forces of demand and supply, and those who do the same by simply looking at a chart. You will read about the powerful London bullion dealers who receive cash for every ounce of gold they sell, and their counterparts in the New York, Chicago and Hong Kong futures markets who deal in billions and yet rarely get to see the gleam of the metals they buy. And finally, you will learn about the *politics* of precious and strategic metals. After all, those who used them wisely have always ruled our planet — and by the look of it, they always will ...

The Precious Metals 1

Introduction

Precious metals are as old as money. In fact, for a long time they *were* money, and today's living evidence of that is in our language. In French, the word "argent" means silver, but today it is also generally used to describe money. In the German or Dutch language, money is called "Geld", differing from its original meaning by only one letter. The Chinese character for gold is also part of the symbol for silver and, perhaps more important, of the symbol for money. And in most of the Spanish-speaking world, transactions are settled in "pesos" or "pesetas". Both these words describe a weight, as does "pound". Obviously, the original pesos, pesetas and pounds were different weights of precious metals.

Essentially, anything can serve as a sound money if its supply is limited. That is why beads, colourful shells, or the teeth of certain animals provided an efficient means of exchange in the regions in which they were used. However, when men began trading more extensively, they discovered that what was rare in one part of the world was not necessarily in great demand in another. Whole economic systems collapsed as a result, leaving only those materials which were genuinely rare as true currency.

Gold and silver were valued right from the start. They were not perishable or destructible, as were oil, grains or other barter items. They were very dense, so that great wealth could be stored in a relatively small and easily transportable hoard. Above all, they could be crafted into jewelry which, to the rulers of the ancient world, was very important. Like many a millionaire today, they liked to display their wealth.

Gold

THE HISTORY OF GOLD

Gold was produced as long as five thousand years ago, when the Sumerians used it for the creation of ornaments, jewelry and pieces of weaponry. However, some historians think that the discovery and knowledge of the yellow metal can be traced much further back, to between 8000 and 9000 B.C. They believe that the oldest items were melted down and used over and over by succeeding generations. We also know that, in ancient times, gold had magical significance. Priceless artifacts made of gold were used in religious ceremonies and have since been found buried beside the bodies of pharaohs and kings.

Inevitably, gold gained in economic and political importance as well. As early as 3100 B.C., the Egyptian treasury used bullion in the form of bars and wafers as a store of wealth. During the next six hundred years, the metal became increasingly popular in international trade. Hieroglyphic accounts of a trade voyage to the mysterious land of "Punt" depict the glorious return of vessels loaded with large quantities of gold bars and rings, as well as other valuables of the era, such as leopards, monkeys, throwing sticks, ebony, ivory and incense.

---

fact:
In ancient Egypt a skillful gold beater could hammer gold bullion to such fine consistency that it would have taken 250,000 sheets to produce a layer one inch thick. Today, one single ounce of gold can be drawn into a fine wire 50 miles long.

Around 550 B.C., the first gold coin was struck. King Croesus of Lydia, in today's western Turkey, ordered his own image to be stamped on each gold piece, thereby creating a medium of exchange whose value was guaranteed by the ruler. Because Croesus was overthrown by the neighbouring Persians shortly thereafter, his guarantee became irrelevant, but the practical implications of his idea were to live on. A sound basis for the monetary transactions of thousands of years to come had been established.

Less than two hundred years later, the world lay at the feet of Alexander the Great. Not only had he conquered the largest territory any man had ever controlled, but he had also amassed most of the world's gold. In addition to the riches of Egypt, he had all of Persia's gold as well. When Alexander sacked the royal treasure at Susa, two million pounds of gold and silver bullion and 500,000 pounds of gold coin were his for the taking.

The next major episode in the history of gold came with colonialism. The Spaniards came to the New World in search of spices, but when they saw the vast treasures of Aztec gold, they quickly forgot about the former. In their lust to steal the yellow hoard, they massacred over 50,000 Indians.

Thus, the yellow metal fuelled man's drive to discovery and pushed him beyond the then-known geographical limits. In Peru, Pizarro found even greater treasures than Cortez had encountered in Mexico. The Incas referred to gold as "the tears wept by the sun", which they worshipped as their god. When Pizarro saw the immense supplies of bullion and the golden ornaments which adorned their temples, he decided to capture Atahualpa, the king of the Incas. As a ransom for his release, he demanded that the king's quarters be filled with gold nine feet high, but when this was done he deceived the Incas by killing their leader and by plundering an estimated thirteen tons of gold artifacts. In order to facilitate the transport back, each item was melted down and lost to the world forever.

Although many corners of the Earth lay still unexplored, it was not until pioneers arrived in the North American West that more gold was discovered. These finds were so large that, in the first ten years, more gold was produced in California than Spain had collected in a hundred. It all started with the arrival of John August Sutter, who escaped numerous creditors, his wife and his three children in Switzerland by trading his bankrupt business for a steamship ticket to North America. Sutter made his way to the west coast where, with the help of 150 slaves, he tried to farm a property of more than a hundred square miles in the Sacramento Valley. In 1845, he was joined by James Marshall, a carpenter, whom he commissioned with the building of a sawmill.

Three years later, when working on a ditch which was to bring water to the new mill, Marshall discovered some small yellow pebbles amidst the accumulated gravel: it was gold!

Marshall galloped back to Sutter's farmhouse and the two of them decided to keep it a secret, but it was already too late. Word of the discovery had instantly spread among the other workers. During James Marshall's absence, they had found more gold and had rushed to Sam Brannan's nearby store to convert it into goods.

Now, Sam Brannan was a very shrewd man. Knowing that everyone looking for gold had to come past his store, he immediately left for San Francisco to spread the word about the new discoveries, but not before ordering massive supplies of picks, shovels and other equipment which could be sold to the prospectors. After Brannan's electrifying news was carried by the city press, hundreds and thousands took off for the Sacramento Valley. Urged on by the talk of gold and riches, soldiers deserted their units, and workers left their factories. Once they arrived at the hastily put together mining settlements, however, they were usually bitterly disappointed. Supplies were so scarce that an egg cost almost ten dollars, a loaf of bread was often sold for $25 and the fortunate few who could afford to sleep in a bed paid $500 a month for the privilege! Sam Brannan was rich.

Later that year, the news reached the eastern part of the United States and in early 1849 it spread beyond America's national borders. Over 40,000 speculators came from as far away as Australia, the Far East, and Europe. By 1858, 24 million ounces of the yellow metal had been extracted and many a prospector had become a millionaire.

Ironically, John August Sutter was not one of them. Sutter's dream of starting over again by building a farming operation in a peaceful and distant valley had crumbled under the onslaught of thousands of noisy strangers driven by greed. Even his own workers took to the hills to dig for gold and left a fortune in wheat and hides to rot in Sutter's barns. He travelled to Washington to assert his claims in court but, once again, was out of luck. Disillusioned and impoverished, he finally died on the steps of the Capitol.

Thus, the Great Gold Rush broke some men, while it made the fortunes of others. But largely overlooked is the fact that it also made cities and that, when the prospectors settled down, it made California.

---

fact:
Gold is found in a variety of colours: silver and platinum impurities make gold white, traces of copper give the metal a reddish colour and iron produces varying shades of green. One of the rarest forms is black gold, which contains traces of bismuth.

In 1859, the "Comstock Lode" was discovered. Impoverished Irish miners had found gold near Virginia City, Nevada, and sold their claim to a man called Harry Comstock. His claim was soon to become the United States' primary source of gold for the next twenty years, producing over $130 million worth of bullion. A few years later, South Dakota became a new target for prospectors. The largest gold mine in North America, the Homestake, had been found. With an output of more than 300,000 ounces of gold annually, Homestake is still the largest U.S. producer today.

Gold fever lived on throughout the 19th century, and the next episode took on even more massive proportions. This time, 60,000 men fought their way to the most unfriendly territory imaginable: The Klondike. Large portions of Alaska and Canada's north were developed in the process and over 4,000 men struck gold.

But, in the meantime, major discoveries were also being made outside of North America. An unsuccessful "49-er", Edward Hargraves, had returned to his native Australia in 1851. For some reason, he decided to continue his search for gold and his first find created the largest influx of prospectors in history. Australia's population grew from 400,000 to 1,200,000 in a decade.

---

fact:
the largest gold nugget was found in Australia in 1872: it weighed almost 3,000 troy ounces or over 200 pounds and was named "Welcome Stranger" by its lucky finder.

In South Africa, the Witwatersrand Reef was discovered in 1886. Nearly 300 miles long, it could not be mined using the technology of the time. It took almost fifty years, and extraordinary amounts of operating capital, before the Reef's potential could begin to be exploited. Today, the South African gold mining companies produce approximately 700 metric tons of gold a year. But, in order to obtain one single ounce of fine gold, an average of three tons of ore must be extracted and processed! From Ancient Egypt, where gold was mined laboriously by hand, production has advanced to a complicated system using subterranean pumps, turbines, drills and even locomotives.

But the past two centuries have not only brought fundamental changes in the supply and production of gold. As technology moved on, new uses and applications were constantly found, making gold an invaluable industrial commodity. Yet gold never lost its mystique: the metal that is the first element to be mentioned in Genesis also accompanied the first human on his walk in outer space, covering the umbilical cord which linked him to his craft.

Equally little has changed in the public perception of gold as the basis for a sound monetary system. When soldiers in Ancient Rome and Greece wanted to be paid in gold or silver coin and not in some other metal whose supply was unlimited, they basically distrusted the rulers of their time. Today, such suspicion still exists, although for different reasons.

Fine Gold Production Through the Ages	Metric tons
16th century	36
17th century	45
18th century	90
19th century	4,864
20th century (to 1983)	84,104

A brief glimpse into recent British and American monetary history will illustrate the reasons for this distrust in more detail. In 1792, the United States became the first modern nation to formally establish a link between paper money and precious metals. The U.S. system was a "bimetallic" one, consisting of gold and silver. Great Britain followed shortly after, basing its money exclusively on gold. Coins and bars in circulation could be freely exchanged against paper currency issued by the Bank of England, and vice versa. Thereafter the use of paper money outmoded the metal's role as an exchange instrument between individuals, but every pound in circulation was fully covered by a sufficient amount of gold in the central bank's coffers. And because several governments soon adopted this system, gold evolved into a widely recognized vehicle for international trade settlements.

The First World War caused the creation of substantial debt and, predictably, inflation followed. Between its end in 1918 and the early 1930's, most countries were forced to abandon the convertible currency system. The value of gold experienced drastic increases and that of paper money suffered. Finally, the world's two most powerful currencies were also disassociated from gold: the pound was devalued by 30% in 1931, and the dollar by 41% two years later!

World War II threw the system into even deeper chaos and, when the war came to an end, there was a universal desire for stability, both politically and economically. The western nations decided to fully reintegrate gold into their money system and established the "Bretton Woods Agreement". All the world's currency values were fixed in terms of the U.S. dollar and, in turn, thirty-five dollars were freely exchangeable for one ounce of gold.

The workability of the Bretton Woods system rested on one of the soundest economic principles: balance of payments discipline. According to this principle, no participating nation could "cheat" by expanding its monetary base without a corresponding increase in productivity. If the United States, for instance, did so, more dollars would chase the same amount of goods and, since some of these goods would be purchased from abroad, more dollars would find their way into the pockets of foreign individuals and companies and, before long, into the coffers of a foreign central bank. Under the dictates of the Bretton Woods Agreement, this foreign central bank could then present its accumulated U.S. dollars for conversion into gold. And since the supply of gold at the U.S. Treasury was limited, the U.S. Government obviously had to be careful not to create too many dollars which could find their way abroad. In other words, the gold reserve of each nation was designed to impose a limit to inflation.

In reality, economic treaties of this kind do not have any greater chance of survival than political ones. Although the catharsis which takes place when the tragedy of a major war comes to an end inevitably nurtures the desire to pool knowledge and resources, the independent economic developments of each nation tend to pull such alliances apart a few years later. Priorities change. Each nation grows on its own and develops its own problems and, eventually, a recession in one country coincides with a boom period in another.

The Bretton Woods system was a typical illustration of this phenomenon. Everyone in the free West wanted the United States to be the most powerful nation and everyone sensed that only America would have the clout and the know-how to reconstruct a Europe and a Japan which had lost their ability to function industrially and socially.

It was in this way that the Bretton Woods Agreement at first played an integral part in post-war reconstruction. Billions of dollars found their way into Europe and Japan where they fuelled a gradual industrial recovery. Foreign governments, which accumulated these dollars, knew that they could convert them into gold at any time, but they did not do so because the dollar was still seen as the world's strongest currency.

This perception started to change in the early Sixties. Eager to maintain a profile of strength and power, America had engaged in the Vietnam War, which had a devastating impact on its deficit. At home, meanwhile, American politicians were busy experimenting with Keynesian economics, whose misguided teachings provided them with the idea that a sagging economy should be spurred on by the creation of even further debt. Predictably, this approach was not successful, largely because the Bretton Woods system worked even better than anticipated. European central banks, already accumulating massive amounts of newly created dollars and anticipating still more, started to convert these to gold. By 1968, the U.S. official reserve amounted to only 296 million ounces, compared to over 700 million ounces twenty years earlier. Observers inside the U.S. and abroad predicted a massive devaluation of the dollar and a substantial increase in the price of gold, to which investors everywhere reacted. Ironically, Americans themselves could not respond because President Roosevelt had banned private gold ownership during the days of the Great Depression.

Escalation of the Official Gold Price		
1792	$17.92	U.S. introduced bimetallic standard.
1834	$20.67	Technical adjustment.
1934	$35.00	Roosevelt bans ownership and devalues dollar by 41%.
1972	$38.00	Nixon withdraws from Bretton Woods Agreement, dollar comes under pressure and is devalued twice. Free market gold price explodes.
1973	$42.22	Dollar crisis continues, but official price remains unchanged to this day.

The western governments spent several years trying to fight the inevitable explosion in the price of gold. At first, they formed the "London Gold Pool", an agreement designed to stabilize the bullion price at $35.00 per ounce. Three billion dollars were spent before speculative pressures became so unbearable that, in March 1968, the gold pool had to be suspended. It was replaced by a two-tier system in which the market was split into an "official" segment, and a "free" segment. The official market, with a fixed price, served for government settlements and reserve evaluations. The free market, in which the gold price freely floated with supply and demand, became the basis for all private gold transactions.

In the meantime, the dramatic decline in the U.S. gold reserves caused a frantic rush among central banks and private holders trying to convert their dollars into gold. Richard Nixon was finally forced to cancel the Bretton Woods Agreement by discontinuing gold's convertibility in 1971. The U.S. dollar, the world's mightiest currency, was no longer backed by gold in any fashion. It was this event which triggered the major decline in the dollar's value which was to last for the next ten years.

Once gold prices were set free, they rose very quickly. Almost immediately, the yellow metal tripled in value, crossing the $100 mark in 1973. And after a brief period of consolidation the climb continued. Once again, inflation rates in the United States, and this time also abroad, were sharply on the rise. Moreover, it became clear that U.S. citizens, barred from owning gold for forty years, would again be able to own the metal. Dealers and private investors speculated that this would bring to the gold market an enormous new source of demand and bought heavily. Ironically, gold peaked almost the same day that the gold market was opened up to American citizens. The price had reached $200 per ounce.

Meanwhile, the frightening developments on the inflation front were being counteracted with high interest rates. America, and most other countries, moved into a sharp recession which lasted through most of 1975 and early 1976. High unemployment and shocking bankruptcies created a bleak economic landscape. Inflation ceased to be a major concern and the price of gold fell, as did other prices.

Washington's policymakers wanted to keep things that way, at least as far as the bullion price was concerned. Afraid of a new surge in speculative demand for the yellow metal, they started to lobby for the "demonetization" of gold. The U.S. Treasury announced that it would auction off a portion of its official stockpile. The IMF did the same but, in addition, urged member countries to use a new accounting unit, the Special Drawing Right, to replace the way they had previously used gold bullion.

The "demonetization" concept turned out to be another disaster. When the U.S. and IMF gold started to come to the market, the price was only a short way from its low of $102, which it had reached in late 1976. But, because the American Government ended its recession by massive creation of new debt, the gold auctions were always over-subscribed. The bullion price rebounded and, as inflation once again became the public foe, soon exceeded its previous peak of $200. In

1979, both the United States and the IMF were forced to reduce, and later to abandon, their gold auction program. The International Monetary Fund's new SDR unit fared no better. Special Drawing Rights are still used as an accounting unit today, but they never provided central banks with the liquidity and independence gold bullion had given them.

The price rebound of early 1979 was impressive, but it was only the modest beginning of an advance which was to carry gold to heights beyond anyone's imagination. By summer, inflation was gaining ever more rapidly and the debt, its major cause, grew at an alarming rate. It had taken decades to create a debt which, in 1970, amounted to $400 billion. Now, only nine years later, the figure had doubled and was rapidly approaching the one trillion dollar mark! To make things worse, several political conflicts of major proportions started to erupt. The worst were those in the Middle East which threatened to interrupt the flow of oil to the western economies.

When the Government of Iran fell to the revolutionaries behind the Ayatollah Khomeini, this was seen as a very significant event. Investors no longer only feared hyperinflation, they were now also concerned about a major shift in the balance of power between Soviet Russia and the United States. That is why, when the Russians invaded Afghanistan in January of 1980, the stampede into gold drove its price to the highest it had ever been in history — $852.00!

---

fact:
All the world's gold could be comfortably placed on any modern oil tanker — that is if any insurer could cover its cargo of over one trillion dollars. . . .

Obviously, there had to be a sharp decline once the hysteria subsided. Politically, the new trend began with the election of President Reagan, who confronted Soviet Russia in a firmer way and, through his rearmament program, reassured the world that the balance of power between the U.S.S.R. and America would be maintained. Moreover, the Russians were soon sidetracked by problems in their own orbit, such as the ones in Poland.

On the economic side, a high interest rate policy was again adopted. The cost of money was pushed far ahead of the rate of inflation, thus discouraging speculators and holders of excessive inventories, and rewarding those with large cash positions. The world went into another shocking recession, with the prices of goods and services dropping drastically. As surely as unemployment and bankruptcies increased, the price of gold declined. From close to twenty percent, the U.S. inflation rate fell to around five percent in 1982. The bullion price, in line with this development, tumbled to below $300.00.

But then a new problem emerged. The world economy grew so feeble that many countries which had borrowed excessively during the preceding high interest rate cycle could no longer find markets for their raw materials. Without their normal income from resource exports they encountered increasing difficulties in paying interest on their staggering debt burden which, in turn, raised even greater questions about their ability to ever repay the loans themselves. Such difficulties were not confined to foreign nations — major institutions in Europe, Japan and North America also faltered.

As the economic crisis deepened people began to wonder about the safety of the money in their banks. The result was a general recovery in precious metals prices as it became evident that a worsening of the situation could only lead to a grave monetary crisis, and that an improvement could only be engineered through the creation of massive new debt. Each of these scenarios suggested a period of turmoil in financial markets, precisely the kind of environment in which gold would do well. The bear market of the early 1980's was finished.

Moreover, it has since become clear which of the two economic extremes western governments are opting for. The money supplies of the United States, as well as other industrialized nations, are increasing at a rate quite sufficient to bring back inflation within a very few years. In fact, investors have already begun to react. By the time inflation is again in the headlines, the current uptrend in gold prices will have become a roaring bull market.

THE FUNCTION OF GOLD

When looking at gold's long history, one cannot help wondering why man has been lured into treachery, revolution and conquest in his search for the yellow metal. We are astonished how, invariably, holding gold translates into power, prestige and wealth. What is so fascinating about gold?

There are many answers. The financial analyst would probably point to gold's reliability as a barometer of monetary and economic health. A craftsman working with the metal is impressed by its incomparable qualities. Gold's even distribution among governments, institutions and private holders throughout the world, and its unsurpassed negotiability, are the most important factors to the

$us/troy oz.

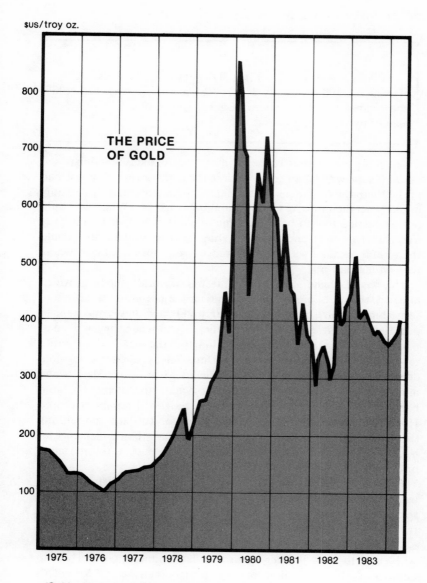

THE PRICE OF GOLD

(Gold price in u.s. dollars per ounce)

dealer. Investors find the metal's rarity a primary asset.

---

fact:

Gold does not rust and is virtually indestructible. It is malleable, ductile and extremely dense. A standard briefcase can hold approximately 4,000 ounces of gold, currently worth approximately two million dollars!

The thing that impresses me the most about gold is best illustrated in an episode which happened back in the early Seventies, at the end of the Vietnam War. Tens of thousands of refugees literally had only hours to get their most essential belongings together. Whatever could not be carried had to be left behind when they boarded hopelessly overcrowded ships and planes. What, in their desperation, could these refugees take with them when they did not even know where they would end up?

The wisest chose gold bullion. It was the only medium which allowed them to condense a significant amount of wealth into something the size of a chocolate bar, and which they could hope to negotiate against cash almost anywhere. And when they arrived at U.S. air force bases, it did not take long for the local authorities to realize what was needed. The State Department called in some gold dealers who, working with simple scales in primitive shacks and tents, purchased gold in exchange for American dollars. The gold bars were foreign, made by refiners in Saigon, Phnom Penh and Hong Kong, and the dealers had to charge for assaying and refining the metals. However, when measured against the convenience of being able to acquire local currency in an emergency, this was a small price to pay.

History is full of such examples, all of which illustrate gold's usefulness as a long term insurance against monetary, political and social uncertainty. The most recent example was the tragic exodus of the "boat people" from Southeast Asia. How did they buy their way out of their countries? With gold. What do you think they carried with them to start all over again in a new country? Gold. How do families in inflation-ridden countries protect their wealth? Where it's allowed, they own gold bullion. And where that is forbidden, their women wear gold jewelry.

Those opposed to a direct relationship between government paper money and gold have often said that the metal has no place in our modern world. They would prefer to see the monetary system tied to our supply of food or to our actual industrial production. The problem with this theory is simply that it is not practical. Such a system would enable governments and corporations to set up huge stockpiles of the chosen commodity, while those individuals without the necessary facilities would have a hard time doing so. After all, storing wheat in a basement, or oil in a backyard, is not everyone's favourite pastime. Industrial commodities are usually bulky and cumbersome to transport. Foodstuffs have the drawback of being perishable. Only gold is not affected by years of storage, is easily hidden and carried, and instantly negotiable anywhere in the world.

Finally, gold's distribution is practically universal. The metal is held by international bodies, governments, banks, corporations, and individuals in almost every part of the globe. It is this simple fact that makes gold so unique and which gives rise to its most important function: unlike other commodities, precisely because it is so well distributed, the price of gold cannot be manipulated.

THE PRICE OF GOLD

If you asked five different analysts to explain why the price of gold moves the way it does, you would probably get five different answers. There are those who argue that gold runs counter-cyclically to stock market trends. Others maintain that gold goes up when the U.S. dollar is down. A third group will pay close attention to the price ratio between gold and silver, while others are convinced that oil, wheat, or other commodities are linked to gold's fluctuations. There is some truth in all of these theories, but none of them have consistency.

The chart on the next page sums up the many factors which, in one way or another, affect the demand or the supply of the yellow metal. Let us start at the top of our table and first concentrate on the supply side, or the producing nations.

Gold Production

Gold producers have fared well during the past few years and they are currently in a state of expansion. Thus, the world's gold supply is gradually rising, a trend which is expected to continue. However, the increases from year to year are not significant and have so far been easily met by growing demand from investors.

The world's largest producer of gold is South Africa, which accounts for about fifty percent of global output. The country's immense gold fields in the Transvaal and the Orange Free State date back over two billion years.

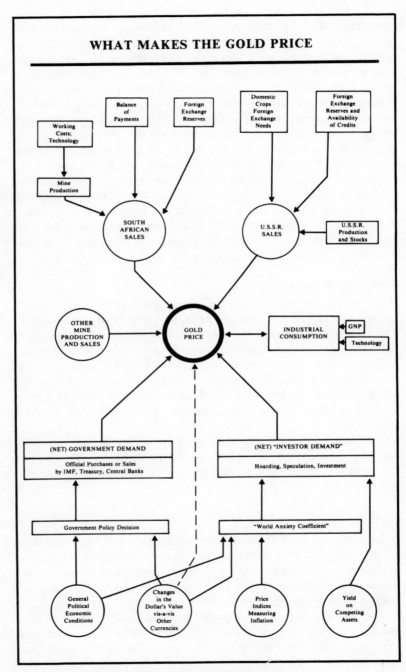

Source: J. Aron, Precious Metals Research Department

Experts believe that glaciers carried mountain debris into an enormous water deposit which was formed in the high velds of Transvaal. Over the next few million years, this inland sea filled up and the land changed. The gold was imbedded in rock and today lies deep below the Earth's surface.

This means that South Africa's mines require the most up-to-date technology. As much as three tons of ore have to be hauled to the surface from depths of more than two miles in order to produce one single ounce of gold. Such difficulties obviously lead to higher production costs although these are somewhat offset by relatively cheap labour.

Less is known about Russia, the world's second largest producer of gold. The first Soviet discoveries go back to the early Twenties, when gold was found in the Aldan River area in Eastern Siberia. News of this find travelled fast and caused a gold rush of similar proportions to that experienced in the United States. Within a mere few months, 12,000 miners had rushed to the site and were looking for their fortune.

Ironically, Russian interest in gold was reactivated by none other than Joseph Stalin. Having familiarized himself in great detail with the economic development of the United States, Stalin concluded that the same principle could be applied to the outlying areas of the Soviet Union. He was interested in the gold of course, but what especially appealed to him was the fact that the American gold rushes had developed whole regions as well. "At the beginning, we will mine gold, then gradually change over to the mining and working of other minerals, such as coal and iron," Stalin wrote.

Today, the Russians are responsible for about thirty percent of global supply. Soviet gold fields are situated in Siberia, Kazakhstan and the Urals, but, of these, by far the most productive operations are those in Eastern Siberia. The geography there is very similar to that of the Klondike or the Yukon and, like the gold mines in our north, Russia's are mostly alluvial. Weather conditions are extreme and permafrost makes production difficult and expensive. However, rising gold prices have helped the Soviets increase their supplies steadily.

Although Canada produces less than five percent of global output, it is the world's third largest producer. Its mining industry has been plagued by declining ore reserves and high costs ever since World War II. In 1941, Canada produced over 170 tons of gold, badly needed to help pay for military supplies. In the following three decades, the mining industry was heavily subsidized by government. The number of operating mines declined from 125 at the end of the war to around 30 in the early Seventies, while output gradually fell back to levels of between 50 and 60 tons. Fortunately, gold's recent price strength has improved the outlook for Canada's mining industry considerably. Not only have the major producing firms shown consistently good profits, but a number of new companies have been formed to pursue exploration projects or, even better, have gone into production themselves. In 1982, fifteen new mines joined the ranks of producers and several more are expected to come on stream during the next two or three years. The most closely watched will be those operating in the Hemlo area of Northern Ontario, where most experts agree that the find is large enough to have a major impact on future Canadian production totals.

The United States, Brazil, the Philippines, Colombia, Australia, Papua/New Guinea and others follow behind Canada. The most promising of these is Brazil where most gold is found in alluvial deposits deep in the Amazonian jungle. Brazil's government recently estimated an annual output of more than 300 tons by the late 1980's. These projections seem unrealistic but, even if only half that target were achieved, Brazil would bypass Canada as the world's third largest producer and would become responsible for roughly 12.5 percent of global supply!

It is interesting to note how politics affect the overall strategies and production policies of gold producing nations. In Brazil and Colombia, for example, large foreign debt makes it attractive for both countries to purchase a substantial part of their own output. Their own gold can be purchased in cheap local currency which, to the outside world, is not of much value. The gold is then held in the two countries' monetary reserves where it serves as collateral to arrange foreign loans.

South Africa's interests have recently been best served by steadily reducing its mining output. From nearly 1,000 metric tons a decade earlier, production had fallen to around 658 tons by 1981. This reflects a shift to exploit lower ore grades, a strategy which is more fully explained in our chapter on mining shares. By concentrating on the production of ores containing the least amount of gold available in the mine, each firm effectively prolongs the overall life-span of its property. And because expenditures are highest when exploiting a low grade ore body, the relatively advantageous labour costs of today are fully utilized.

Finally, it is also in South Africa's interest to have overall global gold supplies follow a steady long-term pattern. By reducing its overall production output while other nations, such as Brazil, are rapidly increasing theirs, an equilibrium is maintained.

WHERE GOLD IS MINED	1980	1981	1982	1983E	1984E	1985E	1986E
South Africa	675	658	664	680	690	700	700
Canada	51	53	63	70	77	85	89
U.S.A.	30	42	43	57	59	77	81
Ghana, Zimbabwe, Zaire, and other Africa	33	38	27	27	27	28	28
Brazil	35	35	35	44	59	80	80
Columbia, Chile, Dominican Republic, Peru and other Central and South America	53	65	60	61	63	65	63
Philippines	22	25	26	25	27	27	29
Japan, India and other Asia	10	10	14	14	14	19	19
Europe	9	9	14	16	16	16	17
Papua/New Guinea, Australia and other Oceania	32	37	46	48	56	69	61
TOTAL NON-COMMUNIST PRODUCTION	950	971	992	1042	1088	1166	1167
Soviet Union, Estimated	260	300	307	312	316	319	324
China, Estimated	40	53	55	57	59	62	67
TOTAL COMMUNIST PRODUCTION, Estimated	300	353	362	369	375	381	391
TOTAL WORLD PRODUCTION, Estimated	1250	1315	1354	1411	1463	1547	1558

World Gold Mine Production, Metric tons

By far the most interesting strategy is that of Russia. Nicholai Lenin predicted that gold would eventually be used to "coat the walls and floors of public lavatories." But he also added that, for the time being, gold presented a big advantage because it commanded a high price and could readily be converted into goods. "When living among wolves, howl like the wolves," he wrote.

During the past two decades of highly volatile gold prices, however, the Soviets, if anything, have been howling with laughter. Unlike the Americans, they consistently used their gold hoard to their best advantage and were by far the most ingenious and disciplined operators in international bullion markets. Of course, it is easier to have an effective policy when a long-term strategy exists. What this strategy is, is not only clear from the Communist manifesto but from hundreds of policy statements released by various Soviet Government agencies since then. The goal of Soviet policy is nothing less than world domination and an integral part of this policy is their strategy to control the world's resources. Gold may be only one of the metals which are predominantly controlled by the Soviet Union and South Africa but, as a key component of the world's monetary system, it is also one of the most important. What today's Soviet leaders would do with gold if they ever achieved their goal is not known. But what we can observe is the Soviet Union's keen interest in the affairs of South Africa and their rapidly growing military presence in the southern tip of the African continent.

In the meantime, how much of their newly mined gold both the Soviet Union and South Africa bring to the market depends, first and foremost, on the price. Both nations obviously want to sell when the price is right and they stockpile gold when it is under pressure. But a combination of other factors affect their policy as well. Most of these are linked to the level of foreign exchange reserves, the balance of payments, and international trade. During the 1970's, for instance, South Africa had to finance heavy military imports, which led to a ten year low in the country's gold reserves by the end of the decade. Similarly, Russia had to step up sales of gold to finance its imports of wheat from Canada and the United States. In simple terms, the leading gold producers build up their reserves whenever they can afford to, and part with their hoards only when it is economically advantageous or when they are pressured.

Official Gold Holdings

When governments come to the market, either to purchase or to sell, their transactions are termed "official". Official transactions tend to influence the market significantly and yet, in almost all cases, the impact is not lasting. This was best shown when the Americans decided to auction sizeable amounts of gold during the late 1970's. At the same time, a U.S. inspired lobby in the International Monetary Fund convinced that agency to follow a similar policy. The immediate impact of such massive new supplies was a drop in the gold price. But, before long, it was recognized that the more gold the United States and the IMF sold through these auctions, the less gold they could bring to the market later. This reversal in market psychology heavily contributed to gold's strength in late 1976 and throughout the following two years. In early 1979 the government formally declared that it would support the dollar and increase gold sales further, but this time the effect on the bullion price was almost nil. And when, half a year later, the Americans had to finally withdraw from their auction program, gold rallied sharply on that news alone.

The Supply of Gold	1979	1980	1981	1982	1983E
Free World Mine Production	959	950	971	992	1042
Known Sales from Communist Block	199	90	280	208	72
Total Supply from Producers	1158	1040	1251	1200	1114
Net Official Sales (Purchases)	544	(230)	(140)	(31)	50
Net Supply to Non-Government Sources	1702	810	1111	1169	1164

Total gold supply, considering new free world production, as well as Communist sales. In metric tons.

But this is not to say that official purchases and sales could not have a lasting impact on gold prices in the future. As the table on page 38 shows, substantial amounts of gold are still held by the western industrialized nations, particularly by the United States. In order to put these holdings into proper perspective, you should compare them to the total of global gold production, which is about 1,400 tons. You will note that the American Government still holds more than five times that figure!

There will probably always be cases where a government, in an emergency, is forced to sell its bullion, as happened with Iran and

Iraq when they recently tried to finance their war effort, and the U.S.S.R. when it desperately needed cash to service its debt to the West. And there will always be other nations whô will want to add to their reserves, as did some of the producing countries in South America and Africa during 1981 and 1982. Colombia, Brazil, Zimbabwe and Zaire all purchased domestically produced gold in order to improve their borrowing potential in international money markets. In addition, several oil producing nations, such as Indonesia, Libya and four other Arab countries have been official buyers for years.

WHO HAS THE GOLD?		
Country	Millions of Troy Ounces	Metric Tons
United States of America	263.66	8,201
West Germany	95.18	2,960
Switzerland	83.28	2,590
France	81.85	2,546
Italy	66.67	2,074
Netherlands	43.94	1,367
Belgium	34.18	1,063
Japan	24.23	754
Canada	20.21	629
Great Britain	19.01	591

Nevertheless, it is unlikely that official purchases and sales will again be a major factor behind gold prices in the near future. What the major nations, particularly the United States, now want is stability. During the late Sixties and Seventies they learned that higher deficits lead to higher inflation, which leads to increased distrust of government and increased investor demand for gold. In 1980, when President Reagan came to power, he recognized these problems and appointed the "U.S. Gold Commission", giving it a mandate to "assess ... the role of gold in domestic and international monetary systems." The Commission studied potential medallion and bullion coin sales by the U.S. Treasury, the issue of gold-backed bonds or notes, the size of the gold stock and the restoration of a gold standard. It concluded that, for the time being, there should be no change in the role of gold, thus brushing aside immediate hopes for a return to a gold based monetary system. Concerning the size of the U.S. gold stock, it found that some limited depletion of reserves would be acceptable, and that the issuance of coins should be considered. These statements all suggest that, on the official front, things will be quiet for some time.

This is not surprising. Once it becomes a component of a country's monetary reserves, bullion represents international liquidity of the first order. It can be held by central banks at superficially low "official prices", giving added flexibility. It also improves a country's borrowing potential and, most important, it cannot be debased or devalued by another country.

Industrial Gold Demand

So far, we have been preoccupied with the role played by the world's governments and with the supplies they bring to the market. It is now time that we looked at the demand for gold, which basically stems from two sources: private investment demand and industrial demand.

Gold is today widely used in a variety of industries, of which the jewelry sector is the largest. Applications in the areas of electronics, dentistry and coinage are also important. The table below illustrates how industrial sources of demand and private bullion purchase each impact on the gold market. As you will notice, "total fabrication", the sum of all industrial demand factors, outweighs private bullion demand to quite an extent. Jewelry is not only the largest item in this group, but also the most interesting. Note how periods of low gold prices inevitably bring about an increase in the demand for jewelry, while sharp rises in the gold price tend to dampen it. While jewelry demand exceeded 1,000 metric tons in both 1977 and 1978, as a result of the rising gold price in the following two years the total dropped to just above 100 metric tons! As precious metals prices fell, in 1981, demand for jewelry rose again, and in 1982 a further increase was posted.

Such a brisk recovery in the demand for jewelry is often an early sign of an emerging bull market. This is because, oddly enough, the most successful gold investors are in the third world, where hoarding and dishoarding jewelry are the only ways to exploit the inflation cycle. In 1974 and 1980, for instance, significant amounts of hoarded jewelry were melted down as investors sold their gold at top prices. On the other hand, jewelry demand for developing nations increased in 1970, 1976 and again in 1982. Historically, the turnaround in jewelry demand has proven to be a reliable signal that the market has bottomed out.

The Demand for Gold	1978	1979	1980	1981	1982	1983E
Jewelry	1008	738	123	599	716	690
Electronics	93	109	100	96	102	110
Dentistry	90	87	62	68	62	50
Other Industrial	68	65	50	47	34	30
Medals & Medallions	50	34	37	28	40	30
Official Coins	286	289	239	243	168	190
Total Fabrication Demand	1595	1322	611	1081	1122	1100
Private Bullion Purchases (Sales)		144	379	262	(56)	53	50
Net Demand from Unofficial Transactions	1739	1701	873	1025	1175	1150

Industrial demand for gold and net changes in investment holdings. In metric tons.

Private Gold Investment

When considering the stockpiles held by producing nations, the vast supplies of gold stored in central bank vaults and the influence of major industrial users of the metal, it is hard to believe that the private gold investor can affect prices at all. And yet the combined thinking of millions of hoarders, traders and speculators who are not the least interested in gold's industrial potential or in its production outlook, can cause far greater price fluctuations than any other factor.

This is even more surprising when we take into account the relatively small figure which private investment demand represents, both in comparison to other demand segments, as well as when measured against the much larger supplies of gold. However, you must remember that the statistics we have so far examined only concern themselves with physical gold. A great number of investors, speculators and traders don't buy gold in the form of bullion, but resort to mechanisms such as the futures exchanges, where the monthly volumes are often larger than the annual production output!

We will take a closer look at this subject in our chapter on futures contracts and will for now concentrate on what causes private investment demand to grow and to ebb off.

Changes in investor thinking are often easy to anticipate, but their impact on the gold market is not. Gold is such a sensitive barometer of the health of our economy that it reacts to changes earlier than most other indicators. As a result, a good deal of the gold price depends on how large groups of individual investors, traders and speculators interpret each economic event as it comes along. Chances are, they will be wrong as often as they are right and the market, for a short time, will behave illogically.

But gold is not only a barometer of our economic health, it is also an important indicator of political stability. While this phenomenon is not difficult to explain, it makes it even harder to anticipate private investment demand. After all, it is often impossible to forecast the outbreak of a political crisis, or even a war.

The following pages will show you how each individual factor affects investor psychology and how, in spite of the difficulties, you can anticipate the price corrections of the future.

Gold and Political Conflict

Why does gold demand soar whenever political turmoil erupts somewhere in the world? To begin with, most political conflicts cause severe economic dislocation. Goods which during normal times are freely available to everyone, are suddenly in shortage; services which people are used to, are suddenly interrupted. Inevitably, this causes prices to rise and creates monetary chaos.

And, to no one's surprise, gold rises in such an environment at an even faster pace. Those living in the affected areas can correctly anticipate massive new deficit spending to finance the war which, in turn, will spur inflation ever further. In addition, they will want to protect their assets from a direct political change or takeover. Obviously, paper money is only a promise made by the government of a country. If there is suddenly the suspicion that the government may not be there a few days later, the paper loses its credibility and the holder will try to convert it into the money of a more secure country or, if he can, he will try to change it into gold. I had an excellent illustration of this phenomenon when the Persian Gulf war erupted, when a friend of mine who worked for a European bank in the Middle East called me every morning to purchase sizeable amounts of bullion. He described how wealthy Gulf residents were literally waiting in line with suitcases stuffed with local cash which they were eager to convert into gold bullion or into dollar deposits abroad. A few years earlier, the Thai coup d'etat in 1976 and the Middle East conflicts of 1977 and 1978 caused similar demand for gold.

The larger the potential political conflict, the more sizeable will be the effect on the price of gold. Both world wars resulted in sharp increases in the bullion price and, more recently, the Soviet invasion of Afghanistan had the same impact. Instead of just a small group of locals, a far larger number of investors from around the world will purchase when crises of such magnitude occur.

Gold and Debt

Most of us have been fortunate enough not to have been affected by a major war or political conflict for decades. Instead, our concerns were largely directed to the economic threats to our freedom and independence, particularly inflation.

Inflation has been around for thousands of years, but it is still a concept which is understood by very few. Whenever inflation occurs, politicians will readily find someone who can be blamed for its existence and, amazingly, the public tends to accept such explanations. The scapegoats of the recent past range from landlords to trade unions, from property owners and speculators to the exotic oil sheiks of the Middle East.

But the real culprit is almost always the excessive creation of money or debt. Inflation is merely the effect. It is only once inflationary conditions exist that the prices of commodities and, later, goods and services rise, and a basis for speculation can be created. Our scapegoats, therefore, are only responding to the malaise of rising prices. They are not causing it.

Gold reacts to inflation earlier than other indicators, and to a larger extent. The reason for this is very simple. Since gold is something which affords a protection against inflation, as soon as investors see a larger increase in the supply of money or the creation of debt than is justified, they anticipate more inflationary developments and buy gold. This usually drives the price up immediately and spells the beginning of a new bull market. But the values of other goods and services will often lag behind because fewer investors follow them closely and because the logical chain of events is longer.

When a government decides to "reflate" its economy, it is usually during a time of very slow economic activity, stifled demand and falling prices. In such an environment the prices of copper or lumber, for instance, will be depressed. And when the money supply is expanded or deficits are incurred to spur on these industries again, it will still take considerable time before the values of these

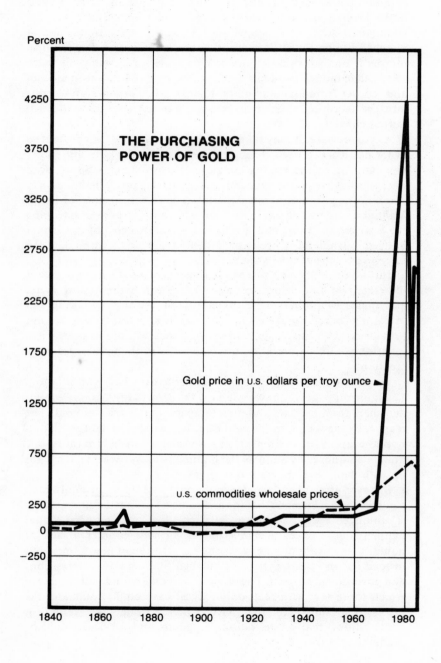

Percent

THE PURCHASING
POWER OF GOLD

Gold price in u.s. dollars per troy ounce ▶

u.s. commodities wholesale prices

commodities are actually affected. First of all, the unemployed need to be brought back to work and consumer spending has to be stimulated. And once people start to spend, they find a lot of merchandise on the shelves which has to be bought before retailers reorder from their supply sources. Wholesalers, too, will want to see steady demand for the items they inventory before they turn around and reorder from manufacturers. Finally, when orders start to pour into the factories, their managers need time to revert to normal output capacity.

As you can see, it may take several months before the consumer's order has worked itself through this system to the pulp and paper merchant or to the mining conglomerate producing copper. And once this has happened, higher prices work back through the chain in reverse. The producer of raw materials is faced with an extraordinary amount of orders and has to find new ways and methods of satisfying these within the time he is given. The additional cost of this is first noticed in wholesale prices which then, in turn, carry through to the price tag on the retailer's shelf.

Just how useful gold's role as a barometer of inflation can be is best illustrated in our chart on page 43. Because the yellow metal anticipates inflationary developments, it has the ability to protect an investor's purchasing power. Most other commodities cannot provide us with this protection because they will rise in price when inflation is already a reality, not when it is in the first stages of its creation.

We concluded that, in most cases, debt is the cause of inflation. The table on page 47 shows you how the growth of the U.S. Federal debt had an immediate and direct impact on the price of gold. The reason, by the way, that the gold price lagged behind debt escalation until the late 1970's is simply that it was artificially held at thirty-five dollars an ounce. As soon as the bullion price was set free, it caught up with debt developments very quickly.

But ever-growing indebtedness not only leads to inflation. In fact, a good case can be made that the end result of such a condition will be a prolonged period of harsh deflation. Just imagine what would happen if, for once, the government did *not* counter a lack of liquidity and economic weakness by the creation of new money, but instead left the entire system of indebted individuals, corporations and government to itself. The obvious outcome would be a climate in which those few with real, unborrowed cash could command the prices best suited to themselves. Before long, the values of commodities, real estate, consumer goods and services would be plunging in a free fall.

Most investors think that gold would do very poorly in such a climate, but that is not true. Surprisingly, an in-depth analysis of the major periods of deflation in recent monetary history reveals that the purchasing power of gold always increased during such episodes. This is not to say that gold did not drop in nominal value; what it means is that gold dropped at a much slower rate than most other items. Next to cash money, it was the best thing to hold.

Thus, gold's ability to maintain its purchasing power during both inflationary and deflationary periods suggests that the yellow metal is the only medium suitable for the protection of assets during both these extremes.

This basic principle is by far the most important thing to remember about gold. Although other factors may interfere with the bullion price for a short time, the major overriding factor determining the price of the future will always be the extent at which debt has been, or is being, created. And yet, the myths that gold moves with the dollar, the stock market, or the oil price continue unabated. Or are they really myths?

Let's take a closer look ....

Gold and the Dollar

"If gold is up, the dollar must be down", goes an adage which held true for most of the Sixties and Seventies. At the time, the United States was plagued by sizeable inflation and deficit spending, while several of the European countries and Japan subscribed to strict fiscal and monetary discipline. Predictably, the dollar was very weak during this period, while currencies such as the Swiss franc, the German mark, the Dutch guilder or the Japanese yen became known as "hard currencies".

During the second half of the Seventies, ever-increasing economic interdependence brought the economic cycles in Europe, the Far East and North America more into line with each other. During 1978, 1979 and 1980, the U.S. was no longer alone in experiencing double digit inflation, and when the U.S. was thrown into deep recession, it pulled the world down with it.

After the advent of the most severe economic slowdown since the 1930's, cash became king again — cash in U.S. currency, that is. The reason, quite simply, was that deposit rates for American dollars commanded such a high premium over the rate of inflation that investors just couldn't resist. The more traditional fundamentals, such as the size of the U.S. deficit, or its poor trade performance, were conveniently forgotten. And as the dollar was high, the level of global interest in gold remained low.

In 1984, this international enthusiasm for the American currency began to wane. Even record high interest rates could no longer mask a steadily deteriorating trade shortfall and an ever-growing budget deficit. Once again, the yen, the franc and the mark posted strong advances and, once again, the scramble for alternatives to the dollar benefited gold.

Gold and the Stock Market

Another relationship which is widely misunderstood is that between the stock market and gold. Most brokers will swear by the formula that a weak stock market must translate into a strong gold price, and vice versa.

First of all, it should be observed that different stock markets react to entirely different factors. Industrially based values, such as the United States stocks, suffer when there is an excessive amount of inflation, while gold obviously benefits. Thus, the Dow Jones Industrial Average was very depressed during the high inflation period of 1978 to 1981. On the other hand, resource-based stock markets in Australia, Canada and South Africa benefited immensely during these same years and were outperformed by very few things, among them gold.

Stock markets are worth keeping an eye on. They reflect investor confidence and the health of an economy. But at the same time, trying to equate stock market behavior with the gold price can be treacherous and misleading.

Gold and Grains

Sooner or later, you will also be told that the bullion price always moves counter-cyclically to the prices of grains. Although gold and agricultural commodities, such as wheat, corn or soybeans would seem to have little to do with each other, the workings of international trade suggest otherwise.

Russia is not only one of the world's primary importers of crops, but is also the second largest producer of gold. And because the Russian rouble is a currency which is not much good outside the Communist Block, the Soviets generally have to pay for their grain imports in gold. (Proof, incidentally, that gold is still the international settlement currency of last resort!) Whenever there is news of a poor Russian harvest, commodities traders anticipate that in the non-Communist world the demand for grains and the supply of gold will both increase. The immediate effect is that grain prices go up while the bullion price weakens.

Of course, the opposite can be true when projections for the

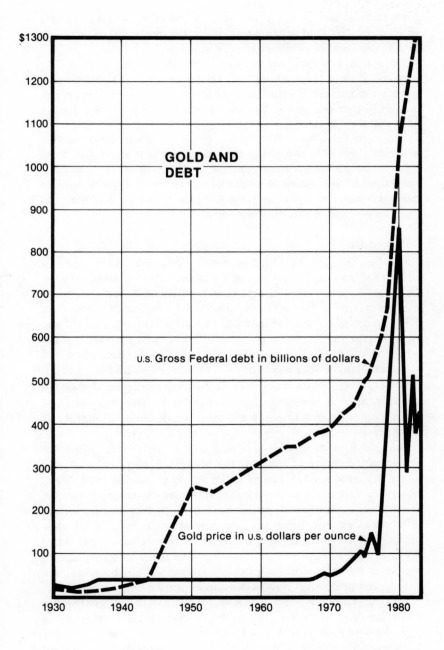

GOLD AND
DEBT

U.S. Gross Federal debt in billions of dollars

Gold price in U.S. dollars per ounce

Russian grain harvest are good. Because less exports to the East are anticipated, less gold is likely to come to the free world in payment. Grain prices suffer, while gold goes up.

Be careful, however, not to rely on this relationship blindly. The Russians do not always hold their gold even when they have had a good grain harvest. Sometimes they have to raise foreign exchange for some other reason and when they do sell their gold, they usually do it discreetly.

You should also know that the Soviet Union's long-term objective to create alternative sources of foreign exchange is starting to show signs of success. Export agreements for Siberian oil and gas have been signed with several Western European nations, giving Russia more flexibility and gradually reducing the importance of the relationship between gold and wheat prices.

Gold and Oil

Finally, we should take a look at the relationship between gold and oil. Throughout the Seventies, there was a direct correlation between the price of crude oil and the demand for bullion. As a result, many analysts started to follow the various oil price indices and, based on these, predicted corrections in the bullion price. As long as oil was the major factor behind inflation, and inflation was the major force behind the gold price, everything worked fine. But when the economy plunged into recession, this principle was no longer valid. In 1982, fear of an international banking collapse dominated investor thinking and gold staged a sharp rally. The prices of crude oil and other commodities, on the other hand, were still sharply on the decline.

But just how dangerous it is to discount such correlations was shown in the spring of 1983. As the collapse in oil prices really got underway, gold suddenly came under selling pressure as well. Many observers were quick to point out that this was no surprise because, as they put it, "Lower oil prices will inevitably translate into lower inflation rates and, therefore, into less demand for gold." Unfortunately, this principle holds true only up to a certain point. Most Western banks chose to base their massive loans to oil producing nations in the third world at prices of around $25 per barrel, which is why that price level now acts as an important support barrier.

What if prices did break that level? Chances are that gold would no longer drop in tandem, but would instead post a dramatic price rise. Western governments would, once again, be caught in a double bind: should they let the banking system get into serious trouble, or should they resort to the printing presses to help the banks and their desperate debtors? We can only guess what the answer would be, but we can say with certainty that investors everywhere would respond by converting their cash into gold.

But whether the outlook for oil is as negative as many analysts predict should be seriously questioned. Remember that the real cause for inflation is the excessive creation of debt by governments. The effect which follows is higher prices for everything, including oil and including gold.

Silver

THE HISTORY OF SILVER

Besides gold, silver is probably the metal that comes to mind most readily when we think about wealth or money. This is easy to understand when we learn that silver was one of the earliest metals known to man. Because it was ductile, very malleable and capable of a high degree of polish, silver was fashioned into ornaments as early as 4000 B.C.

In due course the metal became a medium of exchange. It met the logical criteria of virtual indestructibility and portability and it could easily be divided, as could the other metals which served as money: gold, copper and bronze. Yet silver established itself as the most popular currency, because, unlike gold, it could be found in greater abundance and was distributed over the then known world more equally. In addition, the majority of commercial transactions were too small in value to be paid for in gold, which was very rare and very expensive, and too large to be settled in copper or bronze, both of which were suitable for only very small purchases.

As more silver was mined, more of the metal was accumulated by the rich and powerful. Silver, along with other precious metals, motivated wars and conquests. When Alexander the Great marched upon Persia, he did so not only for its gold but also for its silver. In the ancient world, most silver came from what is today's Turkey as well as from Macedonia and Thrace. Only in the sixteenth century were larger silver deposits found. This time these were in the Americas, where Spanish and Portuguese explorers had discovered them.

---

fact:
The first statistical records of silver go back to the 11th century, when William the Conqueror placed the Mint in the Tower of London and introduced the "Tower pound" as a unit of weight for the white metal.

Through trade, a considerable amount of this silver found its way to England where, under Elizabeth I, an effective silver standard had already been established. Because the country's East India company had to finance immense imports of teas, coffees, spices and textiles, all of which had to be paid for in silver bullion, demand from England was consistently high. The relationship between silver and Asian trade became so obvious that whenever the company prepared a ship for sail, it had an immediate impact on the bullion price! But while large quantities of silver were leaving England, substantial gold imports were occurring at the same time. This was because most of the nation's trading partners in continental Europe paid for British goods in gold. Sir Isaac Newton, then the Master of the Mint, recognized that silver would soon be in shorter and shorter supply and that the price between the two metals was being totally distorted. What evolved was an effective gold standard, although it was not called that until 1816.

In the meantime, sizeable discoveries of silver had been made in the United States. And because the U.S. was also the major producer of gold, a bimetallic monetary standard was chosen in 1792. But the American experience was no better than what the British had gone through: it found itself plagued by massive shifts in international trade which constantly affected bullion prices. Gradually, the U.S. also shifted towards a gold standard, proclaimed in 1900 as the "Gold Standard Act".

As it became clear that more and more nations were switching to the gold standard, substantial quantities of silver began to flow from official holdings onto the free market. Previously, people had been able to come to the Mint and have their silver turned into legal tender, free of charge. In other words, bullion could freely be exchanged into coinage bearing a face value, which gave the white metal a fixed price. The removal of this right to free coinage depressed the price of silver because no one wanted the metal any longer.

Silver production, meanwhile, was sharply on the increase: from about 40 million ounces per year in the 1860's, output rose to 170 million ounces thirty years later. This over-supply was felt the most keenly in the United States which, by then, had become the world's largest silver producing nation. In 1878 the Bland Allison Silver Purchase Act directed the U.S. Treasury to purchase four million dollars worth of silver per month and to coin it into silver dollars. But the pressures discussed earlier led to further price deterioration: having been traded at $1.29 in 1874, silver was now selling at 92 cents.

By 1890, the U.S. was flooded with silver dollars and new legislation, the Sherman Silver Purchase Act, directed the Treasury to increase purchases of the white metal to $4.5 million per month.

Three years later, the government was virtually bankrupt and had to suspend all such purchases. Silver was still dropping. By 1902 the silver price was at 47 cents.

Silver recovered somewhat in the years following World War I, but was then pushed even lower by the deflationary pressures of the Great Depression. From 50 cents in 1929, the metal fell to an all time low of 25 cents in 1933. Nevertheless, with the passing of the Silver Purchase Act of 1934, the Treasury was instructed to buy silver until the traditional Mint price of $1.29 was reached again. A wild speculation to accumulate silver ensued in the open market. Soon the price was back to 80 cents and the government officials expressed surprise and outrage at the profits which were amassed by some market operators. It did not occur to them that they themselves had just created the basis for such speculation and that the original Mint price of $1.29, not seen since the 1870's, had absolutely no relation to reality. At any rate, by the beginning of World War II, the U.S. Government held almost three billion ounces of silver in its coffers!

---

fact:
In the 1890's the U.S. Government went nearly bankrupt trying to keep the silver price high. In the 1960's billions were spent in an attempt to keep the price low. Both efforts failed miserably.

All along, silver producers and industrial users had been the largest beneficiaries of the Government's interventionist policy. While the Treasury had been busy buying silver at prices higher than market value, industry had been quietly purchasing their supplies from abroad, at cheaper rates. In the mid-1950's, the market price started to exceed the price posted by the Treasury, but the principle that the Government should lose on its silver transactions remained intact: silver users now simply purchased their supplies from the Treasury.

During the Sixties, the world economy grew at a tremendous pace, and demand for silver along with it. Eventually, the Treasury's selling price of $1.29 was broken and the market had, all by itself, accomplished what the Government had been trying to do for a quarter century!

One would think that this would have been cause for rejoicing, but now politicians suddenly realized that vast amounts of silver dollars were in circulation, each one of which was valued at $1.29. If they wanted to keep this money in circulation, they had to keep the price of silver *down* or people would simply start to hoard their coins. Thus, a disastrous attempt to keep the price of silver from rising unfolded. It dragged on throughout the Sixties and culminated in the final withdrawal of all silver coinage in 1968. It also left the Government's stockpile of silver nearly depleted: from three billion ounces at the beginning of World War II, the silver stock had shrunk to below 200 million ounces — all sold at below market prices!

Ironically, it was the demonetization of silver and the end of government attempts to manipulate its price in 1968, which led to the most spectacular private market intervention in the metal's long history. This occurred in 1979 when the Hunt family of Texas began aggressively adding to its already sizeable silver futures positions. Along with other factors, the Hunts' purchases caused a significant rise in the price of silver, from around $6 at the beginning of the year to $11 by early September.

We now know that several of the trading firms represented on the New York and Chicago Commodities Exchanges then held short positions in the silver market and were counting on the price of silver going down. But, to their chagrin, the Hunts continued to buy and several other private investors, acting through American brokers and Swiss banks, jumped on the bandwagon. In early December, the March 1980 contract, around which most trading actively centred, crossed the $20 mark. By year end, March silver was trading at almost $30 and exchange officials were in a panic. They had 30,000 short contracts and they had just been forced to put up additional margin of at least $1.5 billion, in one month alone! Besides, the danger of a "squeeze" on March delivery was growing daily more alarming. If the Hunts continued to purchase and the trading firms were forced to add to their short positions, how on earth were they going to deliver all the silver they had sold? Since they obviously could not, it became clear to more and more traders that, unless the price collapsed, they would have to liquidate their short contracts by buying silver — from the Hunts!

Exchange officials were now determined to zero in on the Hunts and other investors. They did this by making several substantial changes in the trading rules. To start with, the number of contracts an individual could own in any one month was limited to 500. Secondly, no more than 2,000 contracts could be held in all delivery months combined and, thirdly, any account with more than 100 contracts was made a "reportable account". These new rules had only one purpose: to force the sale of existing long contracts.

In spite of all this, silver continued to climb. The Hunts and other

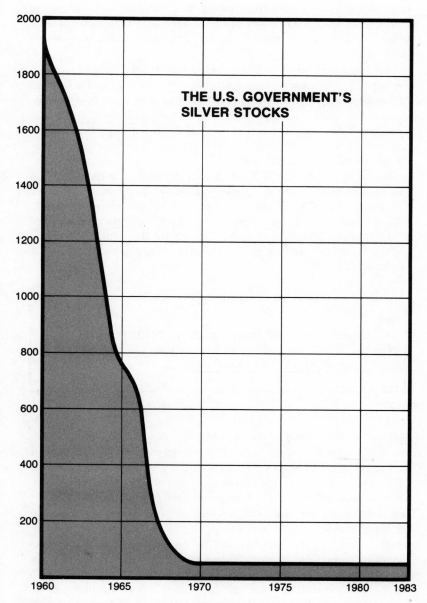

**THE U.S. GOVERNMENT'S
SILVER STOCKS**

u.s. Government silver stocks, in millions of ounces, excluding the official
strategic stockpile of about 140 million ounces.

large investors were now transferring their business to London. On January 17, the price reached $48.80 and the large trading firms were being forced to deposit more than $100 million in margin every single day!

The final bust came on the subsequent trading day, Monday, January 21. The COMEX board actually did what no one thought an exchange could do: they prohibited the buying of silver by anybody — anybody, that is, except those who already held short contracts! Thus, the Hunts could only sell their silver to the same firms which controlled the short contracts. Moreover, they had to sell to them at the prices they bid, when they bid.

Obviously, this was the beginning of the end. The Hunts owned 192 million ounces of silver, bought at an average price of around $10, for a total of $1.92 billion. As of January 17, it was worth $9.2 billion which, if they could have sold it, would have meant a profit of $7.3 billion, or 380 percent. But it also meant that from January 21 on, the Hunts were to lose $192 million for every decline of one dollar in the price of silver!

---

fact:
The Hunt brothers lost more money in the silver market than even the U.S. Government had lost: in less than three months, their fortune shrank by $7.7 billion!

Roughly two months later, on March 27, silver reached its 1980 low of $10.80. The day was named "Silver Thursday". The Hunts had liquidated some of their silver, but their U.S. and London positions still amounted to 141 million ounces, valued now at $1.52 billion. There is no doubt that if the Hunts had been forced to liquidate their remaining silver, the price of the white metal would have plunged to far lower levels.

Still, the price had not reached its cyclical bottom. The global recession gained momentum throughout 1981 and the silver price reflected this. In June 1982, the white metal briefly touched the $5 mark, the lowest price since the days before the Hunt debacle. At such depressed levels, new buying interest set in and silver quickly returned to the more realistic $10 level. And once the first signs of an economic recovery set in, the white metal advanced further. Like gold, silver had entered a new bull market.

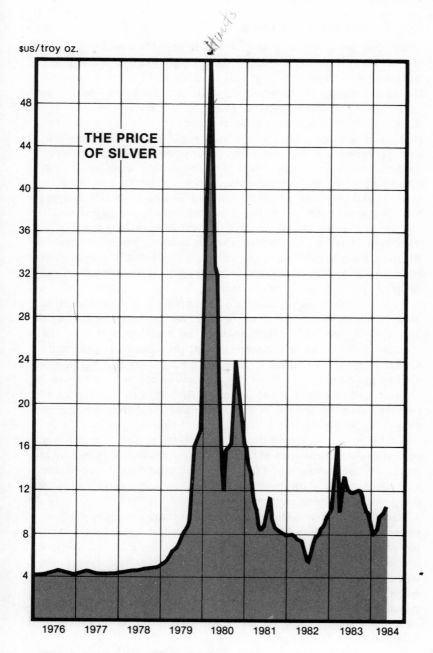

$us/troy oz.

THE PRICE OF SILVER

(Silver price in u.s. dollars per ounce).

THE FUNCTION OF SILVER

Silver, like gold, has a dual function in today's world. To some people the white metal is a refuge to which they can turn when they feel threatened by monetary or economic instability. Others praise it for its incomparable industrial applications. However, there is a difference. The gold price is primarily influenced by investor thinking, while industrial demand is secondary. With silver, it is the other way around. Although most people associate silver with coinage, tableware, decorative objects or photography, more recent applications have taken over. The exploding electronics and computer industries, for instance, use silver because of its high thermal and electrical conductivity, and its ability to reflect light. Small amounts of the white metal are contained in almost any article you use in everday life. Whether it is a watch, a washing machine, a standing lamp or a video screen, there is likely to be a silver rivet, a silver-coated plate, or fine silver wire somewhere in it. This evolution is gradually having an immense impact on the price outlook for silver, which will be discussed later.

Just as most people say that gold has no place in our industrialized modern world, so they wonder how silver can provide investors with insurance against political, economic and social uncertainty. The answer to this is quite straightforward: the desire for protection always stems from suspicion. And when people become suspicious about their money, they inevitably turn to something which is more limited in supply than paper currency. Although less precious and rare than gold, silver is still relatively rare and has therefore managed to protect people's assets over a period of time.

To be sure, many will argue that silver recently did just the opposite: did not millions of people who purchased silver in 1980 at record prices of around $50 lose a fortune? And does this not prove that silver is just the opposite of a protective vehicle, namely a highly dangerous and speculative commodity?

Such questions should serve to remind us that it is only by studying the long term relationship between a precious metal and the pattern of inflation that we can determine its real value. Consider, for instance, the case of someone who purchased silver fifteen years ago, when the price was $1.30, and now looks back on his investment when prices are around $10. On the one hand, our investor has to conclude that he has more than stayed ahead of inflation, but on the other hand he probably wishes he had sold his silver at $50 for an even more spectacular profit.

This example contains a crucial lesson. If speculation is not what you are after, you should strictly follow longer term strategies. But, if you are interested in speculation, then silver will offer you all the volatility you'll ever want — and then some!

Take a look at what happened during the 1979/1980 run-up in prices, when the white metal advanced from below $7 to above $50, thus posting a gain of more than 600 percent. Gold, by comparison, moved from $220 to $850, gaining less than half of that. And when the panic markets of early 1980 collapsed, it was the same all over again. By mid-1982 gold had declined to $280, losing 67 percent of its record value. Silver fell to below $5, giving up more than 90 percent of its peak price!

THE PRICE OF SILVER

Silver and the Economic Cycle

Silver's volatility is easily explained, however, when we look at it in the context of the rollercoaster economic cycle we have been riding for the past fifteen years. Every time inflation needed to be curbed, the cycle was brought to an end through a sharp rise in interest rates. And every time the pressures of recession moved into the foreground and unemployment became the issue of the day, enough debt and money was created to get the economy moving again. At the beginning of each cycle, investors foresaw that industrial demand would sharply increase, but they also sensed that it was just a matter of time before inflation was back in the headlines. Silver catered to those investors in two fundamental ways: not only did the metal benefit from the industrial pick-up but, along with gold, it also promised to protect one's purchasing power. The reverse is true when the economic cycle comes to its peak. Price inflation tends to jack up interest rates which, eventually, stifle economic growth. When that happens, the gold price correctly anticipates an improvement in the inflation outlook and starts to drop. Silver does the same, but it also has to take into account a significant slowdown in the industrial process. Reacting to two such major factors simultaneously, silver invariably drops faster than the yellow metal.

Silver and Market Manipulation

The price of silver has been manipulated more often, and on a greater scale, than the price of any other major commodity. In our chapter on the metal's history, we explored the frequent interventions by the U.S. Government and, more recently, by the Hunt family. Unfortunately, these were only two of the many cases in which the silver price has been influenced by some particular interest, albeit the two most important ones. In other instances, Eastman Kodak, one of the largest silver users, has affected silver and at other times professional commodity traders have pushed the price up or down.

The reason for all this manipulation is really the white metal's inadequate distribution. It is this factor which tempts so many into trying to corner the market. But, as we have seen, things are improving since both the American government and the Hunts have been chastened and since the photographic industry is gradually losing its overwhelming importance as the main source of demand. Still, the white metal has a long way to go before it is as well distributed as gold.

Silver and the Asian Hoard

During the 17th and 18th centuries, immense amounts of silver left Europe for Asia in order to pay for spices, cloth and tea. And to this day the white metal is still regarded there as a store of wealth. You have to realize that most countries in Asia have a "closed" economy with foreign exchange controls and few means of accumulating capital. Silver fits Asian needs beautifully. Particularly in India and Pakistan, the hoarding of precious metals, either in jewelry or in bullion form, is one of the most accepted ways of saving.

During the Seventies, an average of about 45 million ounces of silver were smuggled out of Asia every year. At the beginning and at the end of the decade the amounts were larger, while at the mid-point supplies showed a drop. This illustrates that hoarders in Asia tend to dispose of their holdings when prices are high and reduce their sales when they are low. Although we don't have reliable figures for the early Eighties yet, I am reasonably certain that they will prove to be much lower due to the collapse in the silver price. For the time being supplies from Asia should stay relatively subdued because the silver cycle is one of several years duration and selling is not likely to pick up until we move towards a new top. But, always remember, overhanging any bull silver market, is the huge hoard of the white metal in Asia.

Silver in Exchange Warehouses

Another overhang analysts frequently point to is the vast amount of privately owned silver in the approved warehouses of the American futures exchanges. But what many analysts forget is that these holdings do not belong to the exchanges. Rather, they are largely the property of private and corporate investors who held futures contracts and who then took delivery. Instead of having the metals shipped to them, they simply instructed their brokers to take delivery at an exchange warehouse and have had a receipt issued.

To speak of this hoard as an "overhang" would be a mistake. After all, many of these investors bought their silver at far higher prices than are now available and will want to wait for a price recovery before they even consider selling. The long term trend for private dishoarding of bullion in free markets is exactly the same as in the black markets of Asia. Higher prices cause private investors to dishoard while low prices attract new investment. This was best illustrated when tens of thousands of people lined up to sell their tableware, their tea sets or whatever silver jewelry they could find when silver reached the $50 mark in early 1980. Shortly after this event I went to inspect one of Canada's largest precious metals refineries and what I saw illustrated this point perfectly. A storage room about one quarter the size of a football field was literally filled with drums stacked to the ceiling — each drum packed with cutlery, bracelets and heirlooms waiting to be melted down.

Private supplies held in official exchange warehouses react in accordance with this pattern. They tend to go up moderately during the first part of a bull cycle, then grow at a very quick rate and, following the peak, fall substantially. (The dishoarding *follows* the peak because most investors miss it.) Thus, the danger of renewed sales from this source is over for the time being. In fact, as the silver price starts to appreciate again, warehouse stocks should go up with it thus increasing not the supply of silver, but the demand.

Silver and Soybeans

You may wonder what soybeans have to do with the price of silver and, on many occasions, I have marveled at this curious link myself. Throughout the Seventies, however, there was a relationship between these two commodities.

The reason was more logical and simple than you might think. The whole thing started when a few Chicago grain speculators acquired the habit of converting their profits into silver contracts. When soybeans moved down, they liquidated their silver positions again in order to raise cash to meet their margin requirements. This trend spread and, as soon as an advance in grain prices took place, silver would move up in anticipation of higher prices as well.

Since the early Eighties, though, this pattern has not been reliable. The fall in the price of silver discouraged professional grain dealers from using it as a "parking spot".

Silver Producer Cartels

When the silver price dropped to below $5 in mid-1982, several silver producing nations made strong noises about the necessity of forming a cartel. This was understandable because a prolonged period of such low prices would certainly have forced sharp cutbacks in output and even the closing down of production facilities.

Even so, a silver cartel is a very unlikely affair, either now or in the future. There are many different silver ores in the world, but most of the metal is produced as a by-product of one of the base metals. While some producers mine a particular ore solely for its silver content, to others the white metal is just a windfall by-product. Moreover, production costs vary largely as a result of the different economic structures which exist in today's silver producing nations. The Soviet Union's per ounce cost is quite different from that of the United States, Australia or Canada. And developing nations, such as Mexico, Peru or Chile have foreign exchange earnings to make — even at a loss. In other words, the world's largest producers have different economic objectives. If silver became subject to another selloff, Peru, Chile and Bolivia might very well unite to formulate a policy of their own, but this would be meaningless. As the table of leading producing countries shows, they would control a reasonable portion of the market but other nations such as Australia, Canada or the United States could break their cartel because, together, they command still greater influence.

WORLD SILVER PRODUCTION

Country	1979	1980	1981	1982	1983E	1984E	1985E Estimated
Canada	37	34	36	42	39	40	39
United States	38	32	41	40	43	47	49
Mexico	49	47	53	50	57	65	70
Peru	43	43	47	53	55	57	58
Other Americas	24	24	24	26	27	28	28
U.S.S.R.	50	50	51	50	50	51	52
Other Europe	50	51	49	45	48	48	48
Japan	9	9	9	10	10	10	10
Other Asia	10	10	10	10	10	10	10
South Africa	3	7	7	7	7	7	7
Other Africa	10	9	9	10	10	11	11
Australia	27	25	24	29	33	33	32
Other Oceania	1	1	1	1	1	1	2
WORLD TOTAL PRODUCTION	351	342	361	373	390	408	416

Total world production of silver in millions of troy ounces.

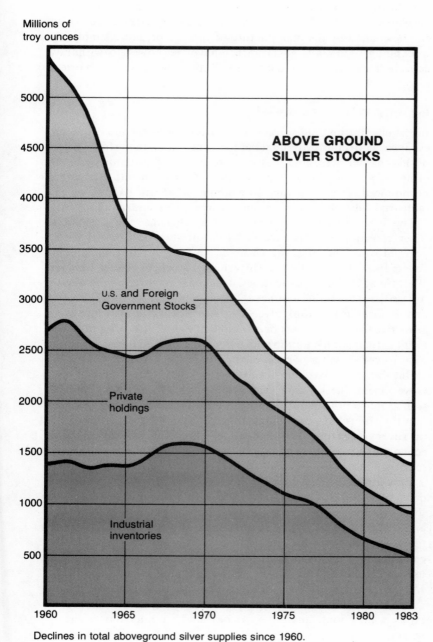

Millions of
troy ounces

**ABOVE GROUND
SILVER STOCKS**

U.S. and Foreign
Government Stocks

Private
holdings

Industrial
inventories

5000
4500
4000
3500
3000
2500
2000
1500
1000
500

1960   1965   1970   1975   1980   1983

Declines in total aboveground silver supplies since 1960.

Industrial producing nations may, however, resort to other measures of price stabilization. In times of depressed prices, demand can be created by the manufacture of coinage or, more simply, by stockpiling. After all, governments have the power to withdraw coinage from circulation and can resell stockpiles again when prices recover.

Industrial Demand for Silver

In the chapter on gold we saw how various political and economic events can affect investor demand for bullion and, in the first section of this chapter, we found that silver reacts to the inflation cycle the same way gold does, except that its movements are more pronounced. It is now time to examine what will happen on the industrial side when an economic recovery gets underway.

This is a rather intriguing subject because, since the late 1950's, silver consumption has exceeded total production in every single year. Most of the silver which made up for this shortfall came of course from the gigantic U.S. Government stockpile which, between World War II and the late Sixties, declined from three billion ounces to less than 200 million ounces. Throughout the Seventies, a lot of private dishoarding took place as prices reached levels which were attractive to even the most demanding investors. And after that, as a result of their confrontation with the American exchanges, the Hunts were forced to sell a portion of their holdings.

But what about the future? Now that the United States Government doesn't have that much silver left, now that we have seen as much private dishoarding as we will see for a while, and now that the Hunts have consolidated their financial position, where will the silver come from? The answer is startling. If we get any kind of an economic upturn, there will simply not be enough of the white metal to satisfy industrial demand!

Photographic materials, by the way, still represent the largest demand factor for silver. Electrical and electronic products are second, while the more conventional applications in jewelry, mirrors and sterling ware are third. As already mentioned, more and more of the world's silver is used in the form of very small, but very vital components of much larger machines. For example, the U.S. military uses more than 5,000 different items containing silver, among them 150 different kinds of bearings. Sizeable amounts of the white metal are used in rockets, torpedoes, jet aircraft, ships, submarines and tanks, while minute quantities can be found in points, starters, gears, valves, batteries, etc.

The amount of silver used in these things is usually worth no more than a few pennies and this means that silver has considerable protection on the upside. Even if the price doubled, tripled or quadrupled, it would have very little impact on the overall unit cost of the many appliances and devices in which it is contained.

But, you may ask, could not the producing nations simply increase supply? Remarkably, the production of silver is extremely inflexible. Remember, the white metal is mined predominantly as a by-product of copper, nickel, lead, zinc and other base metals. The price of silver would have to hit astonishing heights before going after it deliberately would be practical.

The Demand for Silver	1978	1979	1980	1981	1982	1983E
Industrial	449	434	355	337	349	348
Coinage and other	41	28	14	9	13	19
Total Demand	490	462	369	346	362	367
The Supply of Silver						
Mine Production	265	270	255	283	294	310
Deficit covered from above-ground supplies	(225)	(192)	(114)	(63)	(68)	(57)

Demand and supply of silver, in millions of troy ounces. Non-Communist nations.

To generate as much foreign exchange earnings as possible, low-cost producers in the Third World are currently selling all the silver they can. But if you take a look at our table above, you will see that all these supplies can be easily consumed, even though we are only at the beginning of a global recovery. In fact, there is still a handsome shortfall of 57 million ounces which has to be covered from the existing supplies in industrial and private inventories!

Finally, as our figures show, total world consumption for the white metal is again on the rise. If this trend continues, silver will outperform gold in the near future.

Platinum Group Metals

SOME HISTORICAL NOTES

Platinum

Platinum was fabricated in Egypt as long ago as the 7th Century B.C. An ancient casket found at Thebes dates back to that time. Made of an alloy of platinum and other metals, it was dedicated to Queen Shapenapit. More recently, some 400 years ago, the natives of the Upper Amazon made fishhooks and other artifacts from a gold alloy containing platinum.

The discovery of platinum as a separate metal, however, only took place when the Spanish conquered South America. Stream beds in Colombia produced what the Conquistadors called "platina", or "little silver". Colombia remained the only known source for the metal until 1823, when placer deposits in the Russian Urals proved to contain greater quantities. In 1924, the discovery of platinum in South Africa's Merensky Reef surpassed all previous finds and eventually made possible the metal's use in modern industry.

Today, platinum has an importance in our everyday lives that is recognized by very few people — and by a few very astute investors.

Palladium

It was in 1803 that a British scientist found evidence of other metals in a crude sample of platinum ore. One of them was palladium. He named it after a newly discovered asteroid, "Pallas", which itself was named after Pallas Athenae, the protectress of ancient Troy. Also silvery white, palladium looks very much like the element platinum and, like platinum, it is malleable and ductile and has a variety of unusual properties. In spite of this, it was not widely used for industrial purposes until modern times and it has only become available to private investors in recent years.

Other Platinum Group Metals

Iridium, osmium, rhodium and ruthenium all have a great variety of unusual industrial applications. In fact, modern technology could not exist without them. They are scarce, their supply sources are unstable, and their price behaviour is extremely volatile. Unfortunately, none of these metals are suitable for investment because no liquid international market exists at present, nor are there any shares of producing companies which could give you a direct interest.

> **fact:**
> The characteristics of individual platinum group metals vary considerably. Iridium, the "rainbow" metal, produces a brilliant colour-effect when dissolved. Osmium takes its name from "osme", or odor, because it gives off a pungent and toxic smell. Osmium is the heaviest substance found under natural conditions.

THE FUNCTION OF PLATINUM AND PALLADIUM

Both platinum and palladium have the characteristics and, to a lesser degree, the history of a store of wealth. They are rare, precious, have a high density and, in some countries, there are extremely well developed platinum and palladium markets. In Switzerland, for instance, some of the bullion dealing banks often transact more platinum than silver. In Japan, jewelry made out of platinum is at least as popular as gold.

In North America, the investment climate for platinum and palladium is changing rapidly. The refinery and bullion dealing industries have geared up to provide a liquid market for these metals as, increasingly, they are recognized to be a store of wealth like silver and gold.

The functions of platinum or palladium, however, remain largely industrial. If all the world's gold mines closed down tomorrow, the price of bullion would skyrocket but present technology would not suffer much because more than a sufficient amount of gold is already available on the surface. Not so with platinum or palladium. Both metals have become of vital importance to the daily lives of people everywhere. A sharp cutback in their production would force radical industrial changes since, as a recent U.S. research council study points out, these two metals are used as a catalyst in the manufacture of nearly twenty percent of all goods.

In particular, platinum and palladium greatly increase the supply, and reduce the cost, of the world's food and fuel. Virtually all nitric acid — the basis for modern agricultural fertilizers — is made with catalysts using platinum or palladium alloys. As you are probably aware, all North American cars are now equipped with catalytic converters which use substantial quantities of platinum and palladium, but it is not generally realized that most petroleum refining techniques depend on these two metals as well. In addition, they are used extensively in the defense, transportation, communica-

$U.S./troy oz.

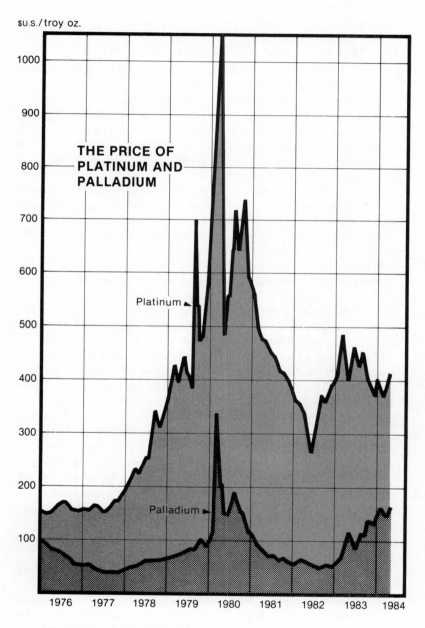

THE PRICE OF PLATINUM AND PALLADIUM

Platinum ►

Palladium ►

(Platinum and palladium prices, in U.S. dollars per ounce).

tions and medical industries. In other words, these metals are of critical importance to our industrial survival.

THE PRICES OF PLATINUM AND PALLADIUM

Like other precious metals, platinum and palladium prices steadily gained throughout the Sixties and Seventies. This was largely attributable to ever increasing inflation and steadily growing political unrest. By early 1980, when both metals reached their peaks, platinum had soared to over $1,000 an ounce while palladium traded near $350. The subsequent global recession weakened industrial demand while high real interest rates discouraged hoarding. The platinum price fell back to $240, while palladium was available for as little as $50. Although both metals have since appreciated considerably, they are still inexpensive when compared with their recent highs. But then, all the precious metals appear inexpensive when compared with the panic prices they commanded in early 1980. So let us use a somewhat more realistic standard.

When I evaluate the *real* value of a commodity, I always compare its market price with its cost of production. Seldom in history has it been possible to find a vital commodity that sells for less than what it costs to dig it out of the ground and bring it to the marketplace. But when you find such a commodity, the chances are that you've hit on an excellent long-term investment opportunity.

As with any other business, a mining operation must generate a profit to survive. When the price of its product falls below the cost of production, there are only two alternatives. One is to stockpile the product and wait for prices to recover (an unlikely option for most companies because patient shareholders are even rarer than precious metals!) The other alternative is to curtail production, perhaps even cutting it back to zero.

Both alternatives have the same effect: they reduce the supply of the commodity and eventually drive up its price. With this in mind, let's look at platinum. The metal's production cost is between $260 and $320 per ounce. In June of 1982, poor industrial demand and fading investor interest brought the price below cost. Predictably, the world's two largest producers of the metal were forced to reduce their output.

From this viewpoint, palladium has been even more interesting. Neither South Africa nor Russia release detailed statistics, but a conservative estimate indicates that their cost of production ranges from $100 to $125 an ounce. This means that palladium, for months and months, was selling at a very attractive discount! Even at current prices, platinum and palladium offer excellent value. As the recovery gains momentum and inflation creeps back into the system, the price of both metals will rapidly gain. Earlier on, I said that silver had a good chance to outperform gold. Platinum and palladium have even greater potential — they could easily become the greatest precious metals investments of this decade!

Precious Metals Markets and Prices

THE MARKETS

Dealers, Brokers and Traders

In the world's free markets, precious metals trading is in the hands of dealers and brokers. The difference between a dealer and a broker is determined by their function. A broker merely acts as an intermediary between a regulated exchange and an investor. To cover his overhead, he charges a commission, but at no time during the transaction does the broker actually own precious metals. A dealer, on the other hand, will purchase or sell a precious metal on his own account in order to make a trading profit. Thus, dealers always have a position in the market, either short or long.

Dealers usually do not charge a commission for the transactions you conduct through them. Instead, they make their money on the "spread", the difference between the price at which they purchase gold and what they can sell it for. Because a dealer lives off his spread, he has to constantly anticipate market movements. If he expects gold to rise, he tries to increase the amount of gold he has and bids a higher price for it than he would ordinarily. Alternatively, if he thinks the market is set for a decline, he may lower his selling price in order to get rid of the position he is sitting on as quickly as possible. Precious metals dealers are usually connected with the banking, investment and refining industries. In North America, the banks and refiners operate primarily in bullion, certificates and coinage, while the investment dealers concentrate on mining shares, futures contracts and options.

A dealing firm's day-to-day business is handled by traders, full time professionals whose job is to identify and exploit market opportunities. In constant search for new information and new markets, these traders are in touch with other dealers all over the world. Quite often, they have full discretion over very sizeable positions. Thus, the dealing community can have an influence on the market all its own, although the basic laws of supply and demand always reassert themselves eventually.

Most bullion dealers derive a substantial portion of their profits from transactions with the public. But an increasing amount of money is also being made from more complicated trading operations, such as "arbitrages". Originally used to describe the simultaneous purchase and sale of a metal through two different dealers or in two different markets, the term "arbitrage" now refers to a wide variety of maneuvers. Dealing in a number of different markets allows a trader to spot the chance of purchasing gold at a lower price in one place than he can sell it for elsewhere. For example, the London market may react downward slightly when a very large gold holding is liquidated there. At the same time, someone else may be purchasing a significant amount of gold in New York. The price difference is usually very small and therefore dealers have to trade in considerable quantities in order to make the transaction worth their while.

More difficult arbitrage operations are those which involve the cost of lending or borrowing money. A trader might notice, for example, that gold for cash delivery is selling at $400. At the same time, bullion for three months delivery is trading at $411. The difference of 2.75% is now compared to current interest rates. At a lending rate of ten percent, the trader can effectively borrow three months working capital at a quarterly rate of 2.5 percent. Consequently, there is room for profit. The trader will therefore borrow as much money as he can at the cost of 2.5 percent. At exactly the same time, he purchases gold in the cash market for the amount he has borrowed and simultaneously sells it three months out. His net profit from this transaction will only be 0.25 percent, but on a large volume that is a sizeable yield. If the trader works for a bank, where the cost of borrowing is considerably lower, the profit is even greater.

But suppose ten traders of large banks exploited the same opportunity. As a result of the increase in demand, the price of cash gold would immediately go up, while the price of gold for three months delivery would come down, reflecting the increased supply. If the three months gold price dipped substantially below the cost of money, the opposite arbitrage possibility would exist. The dealers would sell bullion into the cash market and simultaneously buy it back for three months delivery. The money realized from the cash sale could then be invested in a money market instrument yielding a higher percentage than the transaction cost.

Trading rooms take getting used to. Even a small dealing operation is a place where emotions run high and voice levels rise. Precious metals are a barometer of political and economic well-being, which forces traders to constantly monitor monetary and political events throughout the world. A terrorist raid by African guerillas into Zaire, the latest harvest results from Russia or the release of last month's Krugerrand sales figures are all equally important to the modern gold trader. A North American trading operation begins transacting business with European dealers early in the morning and stays open late into the evening in order to buy and sell precious metals in Hong Kong. But the explosion in communications facilities will probably lead to even further expansion in the future. Some bullion dealers are now toying with the idea of establishing 24-hour trading operations in order to take advantage of all the market shifts and arbitrage opportunities which occur around the world in the span of a day.

Bullion Trading Centres

Most people readily identify the London market as the world's centre for precious metals. Ever since 1666, when King Charles II gave London bullion merchants control over gold and silver dealings, English trading activity has been centred there. Later, when the Empire's large gold discoveries in Australia and South Africa were made, London advanced to the rank of being the world's most important market.

Today, London gold bullion trading is organized by five participants. Mocatta and Goldsmid Ltd., the oldest house, was founded ten years before even the Bank of England opened its doors. The other firms are Samuel Montagu & Co., Sharps Pixley & Co. Ltd., N.M. Rothschild & Sons and Johnson Matthey Bankers Ltd. As they did 150 years ago, so they still meet twice a day for the famous London Gold Fixing sessions. Their representatives get together in a room at N.M. Rothschild's where each has a table with a telephone and a small Union Jack flag. On arrival, each representative determines the amount of gold his institution would ideally like to buy or sell at various prices. When the fixing session gets underway, the chairman calls out a figure and the representatives respond by pointing their flags down or raising them upright, thus indicating whether they would buy or sell at the suggested price. If all attendants wish to sell, the chairman lowers his figure until some show interest in buying. If, at the beginning, they all want to buy, the chairman raises the figure. Once the flags show that there are buyers and sellers among the five, all attendants reveal how much they would like to transact. If the totals are not in agreement, the chairman, traditionally the representative of N.M. Rothschild, makes a proposal to bring the sums into line.

Throughout this session, all five gentlemen are in telephone contact with their institutions' dealing rooms. When the fixing figure is final, it is transmitted throughout the world by every conceivable means. Although only the London five are bound by them, these fixings still serve as an accurate reflection of supply and demand in the world market and are therefore an important indicator to dealers everywhere.

The London gold market functions in accordance with several set trading guidelines. Their dealers specify what bars are acceptable, what fineness they need to have, and where delivery can be made. Most leading gold dealers in North America maintain a "bullion account" with one or more of the five London dealers through which they can settle transactions between themselves. It is quite conceivable that a Far East gold trader and a bank in Canada would agree on London delivery when trading gold with each other.

Still, the London market has declined considerably in importance. In the early Seventies, South Africa shifted a major portion of its gold sales to Zurich, where a newly established "pool" combined the resources of Credit Suisse, Swiss Bank Corporation and Union Bank of Switzerland. Shortly thereafter, when the Soviets brought sizeable amounts of gold to the market, they started to use their own bank which they had established in Zurich to look after the major portion of their sales. Thus, Zurich and London are today of equal importance and, as far as the physical bullion market is concerned, are still the world leaders. Other centres, such as Frankfurt, Luxembourg, Hong Kong, New York, Singapore or Toronto are sizeable enough to provide sufficient liquidity for investors, but rarely affect price trends. Bullion dealers in London, Zurich and these secondary markets, by the way, don't deal in gold alone. In most cases, they are equally active in the silver, platinum and palladium markets.

The Exchange Centres

In 1974, the ban on private gold ownership was lifted in the United States. Banks and dealers opened up trading operations similar to those in London and Zurich, and in time they prospered. But of far more importance was the fact that precious metals could now also be traded on the commodities exchanges. The Chicago Board of Trade and the New York Commodity Exchange correctly anticipated that they could convince investors that buying precious metals with cash was outmoded, and that leverage was the way to go. Within a few years, futures trading changed the structure of international bullion markets forever.

On New York's COMEX, for example, trading in gold bullion soared to over eight million contracts during 1980. This was equivalent to 24,883 metric tons — almost twenty times the annual world production of gold! European bullion dealers found it hard to believe that American views on investing were so different than the trends back home. But, for their own purposes, they too found the futures market to be a very convenient vehicle. If they wished to protect a position overnight, they could simply buy or sell contracts in Chicago or New York and liquidate them again the next day. This possibility was extended even further when Far Eastern markets in Hong Kong and Singapore provided the same facility. North American dealers could now protect their open positions and then offset them in Europe a few hours later. Thus, the leading old world bullion dealers contributed significant volume to the new "paper markets", and precious metals markets literally became a 24-hour affair.

GOLD FUTURES CONTRACT VOLUME

		1975	1977	1979	1981	1983E
COMEX, New York	100 oz.	393,517	981,551	6,541,893	10,373,706	12,300,000
IMM, Chicago	100 oz.	406,968	908,180	3,558,960	2,518,435	995,000
Other Exchanges	33.2-400 oz.	97,936	20,058	311,420	484,209	480,000
Total Contracts		898,421	1,909,789	10,412,273	13,376,350	13,775,000

Source: Consolidated Gold Fields, London

Today, futures markets are even more important. By far the most liquid marketplace, they now serve dealers, producers and industrial users alike. New York's COMEX has become the largest trading centre for gold and silver futures, while the New York Mercantile Exchange dominates platinum and palladium trading. A second tier of exchanges offering precious metals futures include Chicago, London and Hong Kong, among others.

Most observers agree that the world's stock and commodities exchanges will increasingly focus in on precious metals. After all, they are already set up to provide an exchange mechanism, and vehicles such as futures contracts or options have been offered by them for many years. Some experts also fear that this will spell the end of the bullion dealing industry, but they totally forget that investors will always be interested in buying physical bullion as long term insurance. The only significant change will be that both the brokerage community and the bullion dealers will learn from each other, with the result that their knowledge and their servicing ability will improve. We, as investors, can only benefit from this trend.

PLAYING THE RATIOS

Most investors think of their gold, silver or other precious metals in terms of U.S. dollars per ounce. But very few people actually keep track of what the price relationship is between these individual metals over a period of time. Playing the ratios can be a highly rewarding game, particularly for those of you who view precious metals as trading vehicles, not just as insurance. Try to make a habit of asking yourself whether gold is fairly priced in terms of silver, or how many ounces of palladium it takes to purchase an ounce of gold. At best, doing so will alert you to trading opportunities and, at worst, it will make you very aware of the basic economic trends affecting precious metals.

Let's begin by studying the price relationships between precious metals in early 1980, and comparing them to those in mid-1982. As you know, these two time frames are both extremes: the former was a period of record prices while the latter was a period when demand was practically zero. The table opposite illustrates how sharply the nominal prices of precious metals fell, but it is even more interesting to see what happened to the ratios. In early 1980, you could only get fifteen ounces of silver for one ounce of gold, whereas in June of 1982 you could get 58! What this shows us is that the already volatile fluctuations in metals prices are magnified even further in the movement of the ratios. The primary reason for this is that gold's relative fluctuations during the economic cycle are smaller than those of the more industrially used precious metals, such as silver, platinum and palladium. As you can see, this translated into a situation where, at the height of the recent recession, the purchasing power of gold had increased vis-a-vis silver more than three times. The same holds true for platinum and palladium. One ounce of gold bought 0.8 ounces of platinum at the top of the cycle, but in June of 1982 could have purchased more than 1.2 ounces. The same ounce of the yellow metal could purchase only 1.7 ounces of palladium in early 1980, but could later have been traded in for more than seven ounces!

Thus, of all the precious metals, it was again palladium which had the greatest upside potential.

If these were the extremes, are there any norms to which one can expect the precious metals ratio to return under normal conditions? The history of recent years certainly suggests that there are. In a properly operating economy, the ratios should be approximately as follows:

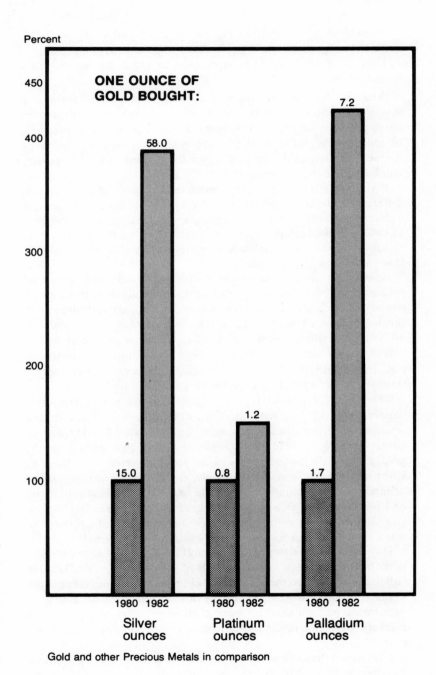

Percent

ONE OUNCE OF GOLD BOUGHT:

Silver
ounces

Platinum
ounces

Palladium
ounces

Gold and other Precious Metals in comparison

1 ounce of gold should be equivalent to 35 ounces of silver.
1 ounce of gold should be equivalent to 0.66 ounces of platinum.
1 ounce of gold should be equivalent to about 2.5 ounces of palladium.

What will be required to create a return to these ideal ratios is a sustained economic recovery. Along with gold, the more industrially based precious metals will anticipate higher inflation and appreciate. In addition, they will also be affected by stronger industrial demand, far more so than gold. The rate of gain for all these metals, therefore, will be higher than that for the yellow metal.

If you wanted to exploit this opportunity, you could utilize two different strategies. The first one is by far the more daring because it involves the futures markets, margin deposits and, if the market goes against you, additional margin calls from your broker. However, futures markets also give you far more leverage and allow you to play this game on a larger scale.

As a ratio trader, you could short a certain number of contracts of gold and go long roughly the same dollar equivalent in silver or palladium contracts. (If this is confusing to you, see the chapter on futures contracts.) Because the timing of such ratio changes is very difficult to predict, you would want to purchase contracts with a delivery date quite far in the future. Your broker, incidentally, would not understand the term ratio trading, but would refer to this transaction as "spreading between two related commodities".

The second, far less risky, strategy you could pursue is that of simply switching from gold into some other precious metal and back into gold. For instance, let's assume you purchased $1,000 worth of silver in January 1974. Let's further assume that each time the gold/silver ratio moved to below 30 you converted your hunk of silver into gold bullion. Alternatively, each time the ratio moved to around 40:1 you exchanged your gold back into silver bullion. If you had followed through and acted on each one of these ratio changes, you would now be sitting on roughly 1,250 ounces of silver worth more than ten times as much as your initial investment!

How would that compare to a straight purchase of either gold or silver over the same period of time? Had you invested your $1,000 in either one of the two metals, you would now have about three times as much. In other words, trading the ratios would have boosted your investment performance by more than three times. In the futures markets, this same result would of course be magnified many times over.

I have used the example of gold and silver to make this illustration, but the situation is exactly the same with platinum and palladium. By

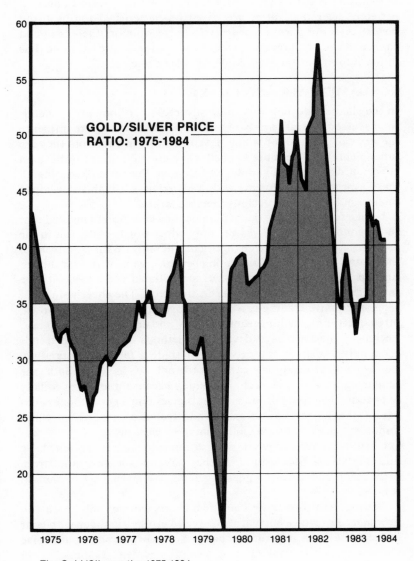

The Gold/Silver ratio: 1975-1984

now, I have probably convinced you that ratio trading is a dead sure game and that by playing one precious metal against another, the risk of exposure is cut down considerably. However, I want to make it perfectly clear that ratios are based on patterns of the past and *history does not always repeat itself*. In my opinion, there is a good chance that we will revert to what I call the ideal ratios at some time in the future, but the wait could be a long one.

FUNDAMENTALS AND CHARTS

In the chapters dealing with gold, silver and platinum group metals we discussed a great variety of factors powerful enough to influence the price of gold, all of which have one thing in common: they are fundamental. To be more explicit, they are guaranteed to have an effect on demand and supply, and thus on the price. Assessing all these aspects is an immense task and requires a highly developed understanding of the precious metals market.

Until a few years ago, the bullion dealing community insisted that such economic analysis was the only effort worth while. But in the last ten years things have changed considerably. With the advent of the futures markets, bullion trading became a game with a far shorter time span. Thousands of contracts are initiated every day, with the sole purpose of liquidation later in the week. The sheer volume of such speculative trading can create short term trends of their own, often overriding the longer term supply and demand fundamentals. Technical chartists have developed methods to read such trends accurately and have thus given their trade a far higher standing. Today, no professional can afford not to look at both the fundamental and the technical side, because they are equally important. The study of fundamentals gives him a good understanding of the longer term trend, while technical analysis provides important clues as to the exact timing of each move.

Obviously, fundamental and technical analysts often have differing views. But fortunately they have also learned a lot from each other and their long-lasting feud has given way to mutual respect.

Getting used to making your own charts will not only teach you how markets behave but will also force you to stay in touch with your investments on a day to day basis. Some brokerage houses give courses on chart making and several newsletters specialize in technical analysis. Of these, I can recommend *Deliberations* and *Aden Analysis*. The addresses for both these publications can be found in our section on newsletters.

Investing in Precious Metals

BULLION

Gold Bullion

The least expensive way of investing in gold is in the form of bullion. North Americans are particularly fortunate because wafers and bars in an extraordinary range of sizes are available to the public. You can purchase gold bullion anywhere from one tenth of an ounce up to the well known "standard bar" of approximately 400 ounces which is used for government and interbank settlements. Moreover, the negotiability of gold bullion is excellent: most dealers can deliver physical gold against cash payment and, much more important, will purchase your bullion back on demand.

Advantages:
- Least expensive way to own gold.
- Instant convertibility into cash.
- Negotiable internationally.
- Direct investment in precious metal.
- Price widely quoted.

Disadvantages:
- Storage risk.
- Money tie-up.

Opportunities and Pitfalls:
- Save on bar charges by buying as large a unit as you can afford.
- Avoid bars made by refiners who are not internationally recognized.
- Try to save sales taxes where possible.
- Deal with an established bank or bullion merchant.

Like all gold, wafers and bars are traded in dollars per troy ounce. Each bar has its weight and its exact fineness stamped on it. Most of the smaller wafers are of a fineness of .9999 ("four nines"), which is generally considered fine bullion. For instance, a 100 ounce gold bar with a fineness of .9999 is classified as exactly 100 ounces of weight and will be priced accordingly. With bullion of a lower purity, this procedure varies somewhat. Because finenesses vary right down to .995 (the minimum accepted standard), the values of such bars are calculated as follows:

Example: When you buy a gold bar stamped 10.0000 troy ounces, and showing a fineness of .9981, at a price of $400, you pay $3,992.40.

(10.0000 ounces x .9981 = 9.9810 ounces x $400 = $3,992.40)

Example: When you sell a kilogram gold bar, stamped 32.1500 troy ounces, and its fineness is expressed as .9990, at a price of $500, you should receive $16,059.00.

(32.1500 ounces x .9990 = 32.1180 ounces x $500 = $16,059.00)

Make sure you deal with a reputable and long established dealer. (A detailed listing appears on page 167.) The main reason for doing business with leading bullion merchants is that they only deal in the products of internationally recognized refiners. This is very important because a bar or wafer manufactured by a refiner who does not have international recognition exposes you to difficulties when you want to resell your gold. If you are not careful, you may have to pay costly assay or melting charges.

Secondly, buying from a leading dealer gives you access to far greater expertise. Most leading institutions have been in the gold business for a long time. But even when you buy from an established bank or precious metals dealer, take a close look at the product you get. Some institutions sell bullion which features their own logo alongside a recognized refiner's hallmark. Stay away from these products also, because your ability to resell the gold later is often limited to the dealer whose name appears on it. And his buying price may not be the best you can get.

Wafers and bars always carry a "manufacturing charge". This means that the dealer passes on to you what it costs him to have your gold manufactured into bullion form. The only item which is exempt from this charge is the "standard bar" of approximately 400 ounces, which is beyond the reach of most private investors anyway.

Manufacturing charges increase the smaller the bar you purchase. The guide below illustrates what charges you should expect when you purchase gold bullion. Obviously, you can save on manufacturing charges by buying as large a unit as you can afford.

Typical Bar Charges for Gold (per ounce)			
1 ounce bar	US$6.00	10 ounce bar	US$2.50
2 ounce bar	US$5.00	1 kilogram bar	US$2.00
5 ounce bar	US$3.00	100 ounce bar	US$1.00

One particular advantage to buying gold bullion is that commissions are relatively low. Some leading dealers do not even have a transaction charge on gold purchases or sales. To Canadians, an added advantage is that most provinces do not levy sales tax on physical bullion. U.S. residents are less fortunate: a few states, such as Delaware, Florida, Nevada and New Jersey don't charge a sales tax, but the great majority of states do.

In 1983, the Internal Revenue Service also introduced sweeping new reporting requirements for gold transactions. Brokers and dealers are now forced to report all sales of bullion and "bullionlike" items directly to the IRS, thus severely limiting the right of individual Americans to keep their gold hoards a private affair.

Canada, on the other hand, is one of the last remaining places where such privacy still exists. Most bullion dealers allow you to settle your purchase against cash, and readily buy gold back from you without the obligation to report transactions.

Silver Bullion

Silver bullion is today fabricated in more sizes than ever before, ranging from one gram (approximately 1/30 of an ounce) to "standard bars" of approximately 1,000 ounces. This aspect, and excellent liquidity, make silver bullion an investor favourite. Many of the points which apply to gold bullion are also valid for silver, although there are some slight differences.

In today's market, all silver is of a fineness of .999 ("three nines"). Wafers and bars of this purity can be resold without assay as long as they have been manufactured by an internationally recognized refiner.

Something else to consider when purchasing silver bullion is the manufacturing charge. Although it costs a refiner roughly the same to make a one ounce silver bar as it does to make a one ounce bar of gold, the same charge impacts on the end price quite differently when expressed as a percentage. Let's assume that gold is selling at $400, while silver is trading at $7. Based on these prices, a $5.00 per ounce surcharge will translate into a premium of only 1.25% on your gold purchase, while it would raise the cost of your silver by over 70%!

Typical Bar Charges for Silver (per ounce)			
1 ounce bar	US$3.50	1 kilogram bar	US$0.50
5 ounce bar	US$1.50	50 ounce bar	US$0.45
10 ounce bar	US$1.00	100 ounce bar	US$0.35

As we saw in the case of gold bullion, silver bars and wafers should only be purchased through a reputable bullion dealer. In most cases, state and provincial tax legislation for silver is identical with that for gold bullion. The obligation of dealers to report your transaction to federal tax authorities is also the same. American brokers have to complete IRS Form 1099 when you purchase silver bullion from them; in Canada, no such requirement exists.

Platinum and Palladium

Until recently, platinum and palladium were only traded in North America in the form of industrial units such as sponge or plate. Obviously, this translated into a tremendous loss of negotiability. In Europe, on the other hand, platinum and palladium have been available in the form of wafers and bars for decades.

This difference is easy to explain. In the United States, investors simply did not know much about precious metals until a few years ago when the ban on private gold ownership was finally lifted. And, in Canada, there was no investor market because banking legislation does not allow chartered banks to purchase and sell platinum and palladium.

However, Johnson Matthey Limited has recently started to produce platinum and palladium in investor sizes ranging from one-tenth of an ounce to 100 ounces. Guardian Trust Company has established itself as the leading Canadian dealer in these metals; U.S. residents can buy them at most Deak-Perera offices. Platinum and palladium bars, by the way, have a fineness of .999 and can be purchased and sold in the same way as gold and silver bullion.

Unfortunately, several disadvantages to physical delivery remain. In Canada, both the provincial and federal governments tax these metals very heavily, leaving investors with the problem that a significant price advance has to occur before a profit can be generated. Provincial sales taxes range from five percent to eleven percent. The federal sales tax is nine percent. In the U.S., state sales taxes are levied as they are on gold and silver, but no federal tax is charged.

Manufacturing charges on platinum and palladium are quite reasonable. Again, they impact more heavily on lower priced palladium when measured in percentage terms.

Typical Bar Charges for Platinum and Palladium			
(per ounce)			
Platinum		Palladium	
1 ounce	US$13.00	1 ounce	US$8.00
5 ounces	US$ 9.00	10 ounces	US$5.00
10 ounces	US$ 7.50		
1 kilogram	US$ 5.00		

BULLION COINS

Gold Bullion Coins

In global terms, gold bullion coins are easily the most popular vehicle for private investors. Best known are the South African Krugerrand, the Canadian Maple Leaf and the Mexican Onza. These three coins are particularly attractive because they each contain exactly one fine troy ounce, which makes it easy to negotiate and evaluate them.

What is a bullion coin and how is its price determined? Governments of gold producing nations make bullion coins in virtually unlimited quantities. The purpose is not to provide circulating money, but simply to create profits. The value of these coins is determined solely by their gold content but there is usually also a small "premium" which represents the manufacturer's cost of production and distribution as well as the dealer's cost of financing his inventory. For all three major bullion coins, the premium is normally between two and six percent.

Let's assume the gold price is $400. The price for a Krugerrand, a Maple Leaf or a Mexican Onza coin should then be between $408 (2% of $400 = $8) and $424 (6% of $400 = $24).

You should know that when you resell a bullion coin you seldom get back the same premium you paid when you purchased. A dealer's buy-back premium is generally far below his selling premium and sometimes there is no premium at all. Thus, your turn-around cost on one of these major bullion coins can be anywhere from zero to six percent.

Unfortunately, that's not where the buck stops. In addition to premiums, there are sales taxes on bullion coins in most U.S. states and Canadian provinces. Because regulations change often, you would be well advised to check out the exact sales tax status with your bullion dealer before you buy. Of course, sales taxes are a direct out-of-pocket expense. When you resell the coin, you don't get back your tax and, to make matters worse, you can't deduct this expense for tax purposes, even though bullion coins are an investment.

Advantages:
— Relatively inexpensive.
— Instant convertibility into cash.
— Negotiable internationally.
— Direct investment in precious metal.
— Bullion coin prices are quoted widely.

Disadvantages:
— Storage risk.
— Money tie-up.
— Premiums.
— Sales tax in most provinces and states.

Opportunities and Pitfalls:
— Compare premiums between coins.
— Fractional sizes cost more.
— Make sure coins are uncirculated.
— Keep an eye on Krugerrand prohibition bill.

Krugerrands, Maple Leafs and Mexican Onzas also come in "fractional units" of one-half ounce, one-quarter ounce and one-tenth ounce coins. These are attractive to investors who cannot afford to purchase one full ounce of gold at a time, and they make very attractive gifts. But, if you have the choice, you would be wise to take the full ounce coin rather than ten one-tenths of an ounce.

This is because the premium on fractional coins is much higher than it is on one ounce units. In the case of the South African Krugerrand, for instance, the selling premiums are as follows:

one-half ounce Krugerrand	6% to 9%
one-quarter ounce Krugerrand	10% to 14%
one-tenth ounce Krugerrand	15% to 20%

As you can see, investing in one-tenth ounce units is not an inexpensive proposition.

There are also several other bullion coins, the most popular of which are:

Austria	100 Corona	gold content	.9802 troy oz.
Hungary	100 Corona	gold content	.9802 troy oz.
Mexico	50 Peso	gold content	1.2057 troy oz.
Russia	"Chervonetz"	gold content	.2488 troy oz.

Although these were the world's most important bullion coins not too long ago, they are of little use to today's practical investor. To start with, their awkward weights make it rather difficult to calculate the value of these coins. Moreover, dealers do not particularly want them in their inventory because they have lost a lot of their liquidity due to the success of the Krugerrand, the Maple Leaf and, to a much lesser degree, the Mexican Onza.

The United States, by the way, is now joining the race with a program all of its own. A series of half ounce and one ounce gold medallions has been minted by the U.S. Treasury, and Goldman Sachs/J. Aron have been appointed as the official distributors. Before buying, make sure there is a reasonably low premium and that good liquidity is guaranteed.

For the time being, however, the South African Krugerrand and the Canadian Maple Leaf are likely to remain the most popular bullion coins, and the choice between them is becoming increasingly difficult. The Krugerrand is better distributed world-wide and, with almost 40 million ounces in circulation, is more negotiable. However, this is starting to change. Introduced just over five years ago, the Maple Leaf is catching up not only in Europe but also in the United States, the Krugerrand's largest market. One reason for this may be that the Maple Leaf is made of four nines pure gold, compared to the Krugerrand's fineness of .9167. But most experts feel that this difference is of no consequence because the South African coin weighs slightly more than one ounce to make up for this variance in purity. Some even go further and say that a lower purity coin cannot be damaged as easily and is therefore preferable.

A far more important consideration is that a bill to ban the importation of Krugerrands was introduced in the U.S. House of Representatives in 1983. Although the bullion dealing industry responded by criticizing such a concept, brokers discontinued their aggressive Krugerrand programs and switched to alternatives. The bill has yet a far way to go, but if it ever became a law, it would seriously affect the liquidity of America's most popular gold coin investment.

For the time being, though, the Krugerrand is still freely available. And when it comes to the most important investment considerations, such as selling premiums, ease of transaction, and negotiability, it is rivaled only by the Maple Leaf.

If you decide to purchase bullion coins, make sure you handle them carefully. You should not scratch them, dent them or nick them. To be on the safe side, it is best to leave them in their original wrapping. You will always get back the basic gold value on your investment but, if the coins are in bad shape when you return them, you may not get any recognition for the premium you have paid. Moreover, because the dealer cannot sell coins once they are damaged, he may deduct a hefty melting charge from your proceeds.

Silver Coinage

Although none of the major silver producing governments has an actual "bullion coin" program, the market itself has filled this void. Fearing the worst, the citizens of Germany, Mexico, Canada and the United States all reacted alike when their governments gradually withdrew silver coinage from circulation and substituted it with worthless base metals. Hundreds of tons of such coinage was privately hoarded and some of it is today still freely traded.

In the United States, anything minted prior to 1965 has a silver content of 90%. In Canada, the story is more complicated but the rule of thumb is that from 1920 to 1966 Canadian coinage contained 80% of the white metal.

In both countries, silver coinage is traded in bulk, usually in "bags" with a face value of $1,000. An American bag of $1,000, by the way, weighs about 720 troy ounces. In Canada, $1,000 worth of pre-1967 coins is equivalent to 580 troy ounces.

Silver bags of this type have several advantages. To start with, they provide the investor with a certain downside protection. If silver ever did drop substantially, you could still go out and freely spend each coin at face value. Thus, bagged silver is both real money and a direct investment in the white metal.

Many analysts also believe that over long periods of time silver coins may appreciate more in value than silver bullion. Particularly in the extreme event of a currency collapse, they argue, silver coins would quickly establish themselves as a favoured medium of exchange.

This is quite true, but investing in silver bags also has disadvantages. One is that they are often hard to negotiate. The interbank market does not deal in silver coinage, because melting it down into bullion is illegal. This leaves investors dependent on the smaller coin dealers who generally run their inventory according to demand. And in the late stages of a bull market this can cause quite a problem. At record prices, when everyone wants to sell, coin dealers will be very reluctant to purchase bagged silver at bullion prices because they will anticipate sitting on the accumulated inventory for some time and will therefore discount the value to correspond with the financing cost.

Selling a portion of a silver bag is even more difficult. When a coin dealer does purchase, let's say, a face value of $300 in silver coins, he will usually charge an even heftier discount. With bullion, on the other hand, you would not run into this problem. If you want to participate in the market to the tune of 600 ounces, you can simply purchase six 100 ounce bars and sell these, one by one, as you see fit.

Finally, there is again the problem of sales taxes. It never fails to amaze me that governments can actually tax the currency of a country, but most take the view that, if the market value is higher than the value stamped on a bullion coin, they have the right to do so.

CERTIFICATES

Gold Certificates

Bullion and bullion coins both have one major drawback: they involve a storage risk. Certificates, on the other hand, provide the investor with a practical means of avoiding this problem and give him a secure alternative.

In most cases, the issuing institution will register your gold bullion or gold coin certificate in your name and hand it over to you. Certificates generally state that you have the right to demand the actual gold you have purchased or, alternatively, its fair market value. In return for keeping the bullion for you, you have to pay a modest storage charge. In addition, most institutions charge a transaction fee at the time of purchase. Gold certificates are now by far the most popular investment vehicles for Canadians purchasing the metal. In the U.S., where they have not been around as long, they are rapidly catching on as a more practical alternative to physical gold holdings.

At least ten dealers issue certificates which, in turn, are sold through at least twenty financial institutions. The Deak-Perera organization, Citibank and Republic National Bank are among the leaders in the United States. In Canada, the Bank of Nova Scotia and the Canadian Imperial Bank of Commerce sell their certificates primarily through their nation-wide branch networks. Guardian

Trust Company, another leading dealer, is represented in major cities and uses most of the Canadian brokerage industry as its sales outlet.

Advantages:
- Inexpensive.
- Highly liquid.
- No storage risk.
- You invest only in a precious metal.
- No sales tax, in most cases.
- Bullion prices are quoted widely.

Disadvantages:
- Money tie-up.
- Most certificate issuers reserve the right to a few days' notice if you want delivery, although this right is rarely exercised.
- Certificates are usually registered in your name — you lose your privacy.

Opportunities and Pitfalls:
- Make sure your certificate is backed by bullion.
- Compare certificate features and charges.
- Avoid conversions into bullion.

Different certificates have different features and different advantages — comparing them carefully can save you a lot of money! The first thing you should examine is the size of the transaction charge. Currently, issuers charge commissions of between 1/4 percent and 3 percent. Of course, it makes sense to pay more than the lowest commission available if you get added convenience out of it. For instance, it may be well worth your while to pay a broker a 2 percent commission if you can do the entire business by phone, when paying a 1/2 percent commission means that you have to run to your bank with your gold in your hands and then stand in line at the bullion dealer's wicket — a time-consuming and frustrating experience if you are trying to sell into a bull market. Nor is the transaction charge the only factor to consider. You will find that some certificate issuers levy manufacturing charges whether you intend to take delivery of the gold or not. Others will only pass this charge on to you if, and when, you decide to take delivery of the physical metal. In the latter case, if you end up cashing the certificate, you save money because you won't have to pay for the manufacturing of bars you never use.

It also pays to check how each issuer's certificates are backed up. Do physical metals stand behind your document or is it secured by more paper in the form of futures or options? Where are the metals held, how are they insured, and how quickly can you get at them? Don't be afraid to ask these questions — after all, you are inquiring about your bullion.

Certificates, by the way, are issued for different denominations. Canada's Bank of Nova Scotia, for instance, recently decided to sell gold certificates for as little as one ounce. The programs of Citibank and Deak-Perera call for a minimum purchase of $1,000. Guardian Trust, on the other hand, insists on a minimum of five ounces of gold per certificate. Most institutions charge storage fees which, at present, are more or less identical. Some dealers also offer certificates drawn on foreign locations. This means that you can take delivery of your gold in market centres outside the United States or Canada. But don't be blinded by meaningless extra features. If you want your gold shipped abroad, most issuers will arrange it for you anyhow, whether it's stated on the certificate or not. And the cost will be just the same.

One final difference between the various certificate programs lies in what type of gold they offer. Most bank programs restrict themselves to bullion only, while firms like Deak-Perera and Guardian Trust also offer Krugerrand, Maple Leaf and Mexican Onza certificates. This is of particular interest to those investors who are discouraged from purchasing bullion coins by state and provincial sales taxes. Depending on where you live, coin certificates give you the opportunity of owning your favourite bullion coin and only having to pay the sales tax if and when you decide to take delivery.

However, if you take delivery, there is one major tax disadvantage to *all* precious metals certificates. Both Revenue Canada and the IRS sometimes consider that a change from a certificate into bullion represents the disposition and repurchase of a commodity and is therefore to be treated as such. More easily explained, if you had purchased a ten ounce certificate when gold was $200, and if you were now to take delivery of that certificate while gold is traded at $400, you would have to declare a $2,000 capital gain — even though you never received a penny!

Obviously, this is the wrong way of looking at the situation because a certificate is nothing more than a receipt for bullion already purchased. But unless you are determined to challenge the tax department on this point, you will have to bear it in mind.

Still, all this only becomes a problem if you take delivery which, of course, is not the purpose behind investing in certificates. The fact remains that, for convenience and ease of transaction, precious metals certificates give you the best value for your money.

Silver Certificates

The same points discussed in the chapter on gold certificates apply to silver certificates. They are issued by the same institutions for sale at their branches or through the brokerage network. Again, it will be worth your while to compare the advantages and drawbacks of individual certificate programs.

The usual minimum for silver certificates is 50 ounces. At present, no major institution issues certificates for silver coins.

Platinum and Palladium

In the U.S., most banking organizations regard platinum and palladium as a speciality area and the market is left almost entirely to the Deak-Perera Group.

In Canada, only Guardian Trust Company offers platinum and palladium certificates. The company's program is represented through most Canadian brokerage firms, allowing you to place orders directly through your account executive.

Certificates are by far the most practical way of owning the physical metal because they overcome the hefty premiums you would otherwise pay in the form of manufacturing charges and sales taxes. Again, it is only if you take delivery that you are exposed to these extra costs.

Platinum certificates are normally issued for a minimum of five ounces, and palladium certificates are available for ten ounces or more. Storage and commission charges are similar to those on gold and silver bullion certificates.

PRECIOUS METALS ACCOUNTS

The late Seventies were a frustrating time for precious metals dealers and investors alike. Prices were extremely volatile and everyone wanted to purchase at the same time, causing long line-ups. For bullion dealers, it was also a very creative time. As a result of the "Gold Rush", dealers expanded their certificate programs to the brokerage industry and brought out a number of new services on their own.

In 1979, one U.S. bank, the First National Bank of Chicago, satisfied their clientele by introducing a gold savings passbook. Operated like a normal savings account, the holder simply buys gold which is then shown in his passbook as a credit balance, or he sells gold to have it deducted from his balance again.

Today, passbook accounts for gold and other precious metals are available from a variety of institutions, some as far away as the Philippines, where Summa International Bank offers the service.

Advantages:
— Convenience and time savings.
— Usually no storage risk.
— Transaction speed.

Disadvantages:
— Usually more expensive to operate.
— Restricted negotiability (you can negotiate only where account is held).
— Money tie-up.

Opportunities and Pitfalls:
— Determine what exactly you want from such an account.
— Compare charges carefully.
— Avoid plans which restrict you, should your objectives change.

A number of programs go a step further and offer additional alternatives. Deak-Perera was the first to announce a program which allows the purchase of gold, gold coin and silver certificates over the telephone. Because the firm does not provide banking services, clients are asked to settle by mailing certified cheques. They receive their certificates later.

A similar service called "Tele-Trade" was introduced by Canada's Guardian Trust. This plan offers nation-wide toll-free dialing and enables investors to buy and sell gold bullion and coins, as well as silver, platinum and palladium. Tele-Trade also pays daily interest on cash balances and allows investors to place free buy and sell orders. One particular advantage is that it does away with such things as manufacturing, storage and insurance charges.

Some precious metals plans are designed to facilitate the *accumulation* of bullion, but are less suitable as a trading vehicle. Among these are the programs operated by Merrill Lynch and Shearson/American Express Inc. Both services allow you to buy gold and silver bullion on an ongoing basis, and in amounts and intervals of your choice. These features are of particular interest to investors who believe in averaging their purchases over a period of time and want someone else to take care of the actual accumulation according to the criteria they choose. One of the most interesting services in this category is "Goldplan", a Swiss concept offering five individualized cost averaging systems, suited for different investment needs. (See our listing of Banks, Brokers and Dealers on page 167.) One of "Goldplan's" programs combines bullion accumulation with optional life insurance, a concept offered also by the Tyndall Group of Bermuda, as well as by several other Swiss firms.

As you can see, each one of these programs has its individual drawbacks and advantages. I suggest that you write to each company in order to compare costs, flexibility and convenience (see our dealer listings, pages 167 to 174). Knowing them well will be increasingly important, because it is in this sector that the largest growth in the precious metals business will take place over the next five to ten years. Running to your bank to get a certified cheque, standing in line at one of the local bullion dealerships and carrying your precious metals home are rapidly becoming things of the past. More and more investors call a trader or broker from the privacy of their home or office in order to place a purchase or sale order. Most of these convenience facilities, by the way, have not been designed with only the aggressive trader in mind, but are just as suitable for longer-term holders of bullion.

With precious metals accounts of this type, it is even more important to "shop around". Try to know what you want from the facility you are looking for before making a decision. If you're trying to accumulate gold with the idea of eventually taking delivery, make absolutely sure that your "account balance" is actually convertible into physical bullion of the size you want. Check, also, whether getting at your gold is practical and doesn't entail unacceptably high shipping or tax expenses. Few of the investors who hold precious metals in bullion accounts in Switzerland, for instance, realize that taking physical delivery would not only bring with it manufacturing charges, but that they would also be taxed with a hefty 5.6% value-added tax!

But perhaps you are not interested in delivery at all and really need an account in order to trade. Obviously, your objective will be quite different. Whether you can place buy and sell orders, what the trading hours are, and how often you receive a statement may be of far more interest.

Finally, you will want to keep transaction charges as low as possible. In the U.S. and Canada, by the way, there are large discrepancies between the commissions charged by individual banks and dealers. They can range from as little as 1% per purchase or sale to as much as 4%, and extra fees such as those for administration, storage and insurance are often added!

FUTURES CONTRACTS

Futures contracts were initially designed to provide protection against fluctuations in the price of commodities and foreign currencies. In the case of precious metals, the people seeking such protection are primarily industrial users. For example, let us assume that a jewelry manufacturer wants to prepare his Christmas sales campaign. It is now mid-summer and during the past three months gold has been trading as low as $430 and as high as $550. Where the gold price will be at Christmas-time is strictly a matter of speculation. At the same time, price lists and catalogues have to be prepared and they of course have to be based on a certain price level. Instead of leaving himself open to further fluctuations, our manufacturer would typically purchase the amount of gold he anticipates will be needed in the futures market.

Advantages:
- You can profit when prices go down.
- Leverage: your money tie-up is substantially reduced.
- Liquidity.
- You invest only in precious metal.
- Futures contracts are widely quoted.
- No storage risk.

Disadvantages:
- Open risk: theoretically, no limit to your losses.
- You are tied to certain contract sizes, maturity dates and trading hours.
- Delivery, if taken, is at an exchange warehouse.

Opportunities and Pitfalls:
- Consider possible tax advantages.
- Study the literature the exchanges provide.
- Start small, build up as your expertise grows.

You can also protect yourself by *selling* on the futures market. Take the case of a mining company as it watches the gold price run ahead of itself and wonders when the inevitable price correction will set in. Smart management will lock in the temporarily high price by selling the gold to be produced during the next few months in the futures market.

But today, those using futures contracts for their protective appeal are not always in the majority. Ironically, futures are frequently used for the opposite purpose, namely for speculation. The main attraction to the speculator is leverage: the ability to put down a small deposit to guarantee the price of a large transaction.

Astonished by the staggering volume of business transacted in North America's commodities exchanges, a European banker recently asked me how many private investors were actually buying and selling futures contracts. When I told him that just about every one of my clients had at one time or another owned a gold futures contract, he looked at me in disbelief. Coming from a culture where precious metals are purchased and then kept for twenty or thirty years, he simply could not understand why ordinary investors would want to indulge in this kind of speculation.

It is true that futures have made many a speculator into a millionaire, but you should realize that for every winner in this game there is also a loser. Leverage can make you more money than you ever expected to make, and it can lose you more than you are now worth!

However, the futures market need not be only for speculators — leverage can be of equal advantage to you even if you are a very conservative investor. Having only a small part of your capital tied up can be invaluable, particularly in a volatile financial environment. Just think of the advantages of having other investment opportunities coming along and being able to utilize them. In short, leverage increases your flexibility to an enormous extent.

From a cost viewpoint, however, futures are no better and no worse than a direct holding of bullion. When you enter a contract, your broker will ask you for a margin deposit, normally between five and ten percent of the contract's value. But this is not the end of your financial commitment: if the market goes against you, you will have to put up additional sums of money in line with your losses. There are also commissions to be considered and, most important, the fact that precious metals for future delivery trade at a premium over the cash price.

The premium for any one metal is usually a direct reflection of the cost of money and, to a lesser degree, what the market expects interest rates and the price of the metal to do in the future. In other words, if you purchased gold for delivery in six months' time, and if present interest rates were twelve per annum, the premium you would have to pay would be somewhere around six percent. If it were seven percent, then you would know that the market expects either gold to increase or interest rates to rise.

One of the key advantages of futures contracts is that they also allow you to benefit when the price of a metal drops. In other words, you can not only buy gold, silver, platinum or palladium for future delivery, you can also sell it. The nice thing is, in order to do so, you don't actually have to own any bullion! You simply sell "short" at a given price and then purchase it back at a lower rate prior to the delivery date stipulated in your contract. And, as long as you liquidate in time, you don't have to deliver the gold either because yours has been a purely paper transaction. But the profits you have made, the difference between the price you sold gold at and the price you paid to repurchase it, are real!

"Going short", by the way, is more speculative than "going long". Since you can lose no more than the total value of your contract, your maximum loss with a long position is limited. If you go short, on the other hand, your risk is open-ended because gold could theoretically soar to unimaginable highs. Luckily, there are some ways to eliminate the danger of losing excessively. Today's futures markets do, in fact, offer a wide range of buy and sell order mechanisms, which can be placed as "stoplosses". Here is how they work:

Let's assume you have "gone short" ten contracts at $473.10. You expect gold to go down and you hope to repurchase it at a lower rate prior to the time you will have to make delivery. But, because you could easily be hurt by rising gold prices, and because you can only afford to lose $10,000.00, you want to place a stoploss. Given that you have sold ten contracts, your risk limit of $10,000 translates into $1,000 per contract. And since each contract is for 100 ounces, $1,000 per contract translates into a net gold move of $10. In other words, you need a stoploss at $483.10.

As soon as your broker confirms to you that he has sold ten gold contracts at $473.10, you would instruct him to place a stoploss for ten contracts at $483.10. Once your order is in, the broker will automatically close your contract out if the market goes the wrong way. If, on the other hand, gold starts to drop as you expected, you can lower your stoploss in accordance with market movements. For example, if gold dropped to $463.10, you could simply cancel your order and put a new one in at $473.10.

If you do this, you should make sure that you place "open orders". These are stoplosses which are effective until they are either cancelled by yourself or executed when the market moves to the price level you specified. "Day orders", on the other hand, expire at the end of each day, leaving you unprotected unless they are renewed at the opening of every trading session.

But stoploss orders, regrettably, don't protect you 100%. Today's futures markets limit commodities as to how far they can rise or drop during any one trading session. Gold's movements, for instance, are limited to a rise or fall of $25 from the opening price. Now let's look at how that can interfere with your stoploss order.

To revert to our example, you have shorted ten gold contracts at $473.10. You feel that you want to limit your losses and place a stoploss at $483.10. During the first few days of your investment things go exactly as planned: the gold price tumbles. But then, in one day, the price moves from $457 to $482, the full $25 permitted during any one trading session. The market closes "limit up" and you think you are safe. But, in the Far East and Europe, there is further buying which pushes the price up another $24 so that, by the next morning when your futures exchange opens, gold is trading at $504! Obviously, your stoploss cannot and will not have been executed. Even if your broker acts immediately, your losses will be much higher than what you wanted them to be.

Another disadvantage of futures contracts is that they are basically "paper gold". It is relatively difficult to take delivery of a futures contract even when the contract expires and you cannot do so at all until it does. One of gold's attractions is obviously that it represents something which is highly portable and which can be exchanged into ready cash just about anywhere on earth on demand. If that is the kind of security you are after, then futures contracts are not for you. But, as we have just seen, there are advantages to futures contracts as well — provided you understand the risk.

Don't be confused by the fact that I have chosen a gold contract to explain the various mechanisms inherent in the futures market. It works exactly the same with silver, platinum or palladium. There are, however, important differences between these precious metals when it comes to liquidity. Even if you held a hundred gold contracts at New York's COMEX, you would never experience any difficulty in getting an immediate bid for these, because tens of thousands of contracts are traded every week. By contrast, when it comes to palladium, for instance, the only exchange at which the metal is traded is the New York Mercantile Exchange, where the business is generated predominantly by industrial users. As a result, it is often

Precious Metals Futures: where they are traded

COMMODITY	EXCHANGE	CONTRACT SIZE (troy ounces)	TYPICAL MARGIN REQUIREMENT	TYPICAL LIMIT MOVEMENT	LIQUIDITY
Gold	IMM, Chicago	100	$5,000.00	$50.00	very good
	COMEX, New York	100	$5,000.00	$25.00	excellent
	Mid-American	33.2	$1,700.00	$50.00	good
	Winnipeg	20	$1,200.00	$30.00	poor
Silver	CBT, Chicago	1,000	$1,200.00	$ 0.50	very good
	COMEX, New York	5,000	$5,600.00	$ 0.50	excellent
	Mid-American	1,000	$1,200.00	$ 0.50	good
Platinum	New York Mercantile	50	$2,500.00	$20.00	good
Palladium	New York Mercantile	100	$1,200.00	$ 6.00	satisfactory

difficult to sell twenty contracts, although fewer than that should be no problem. So liquidity can be an important consideration and, in general, I recommend that you stay with those exchanges where a consistently high volume is traded.

From a taxation viewpoint, futures can be quite complicated to understand, and considerable differences exist between Canada and the United States. Canadian legislation treats precious metals the same as any other commodity which means that, the first time you incur a trading profit or trading loss, you have the right to choose capital gains treatment or income treatment. Once you have done so, however, you have to consistently apply the same standard to all future transactions you may enter into. My recommendation is that you select capital gains treatment, due to the fact that only 50% of a gain is taxable. Unfortunately, not everyone may do so. If you are directly involved in the precious metals markets, or have a professional link with the analysis, investment or trading of commodities, you are forced to declare your profits from precious metals futures as income.

These "insider" restrictions also apply in the United States. For normal investors, however, American laws are much tougher. Capital gains treatment is only available to those who hold on to a contract for a period of more than six months. U.S. legislation has also come down hard on year-end "straddles", maneuvers used by many traders to create artificial losses by selling a contract at an opportune time and simultaneously purchasing it for delivery in the following year. Revenue Canada is expected to announce similar restrictions in the near future.

One final point to consider is that the futures market is a game of timing, not of trends. If you are convinced that gold is going to go up within the next six months, a futures contract may affect you quite differently than a physical purchase of bullion. If you buy wafers, bars or certificates totalling 100 ounces, and the price, at the end of six months, is down and not up, this may not bother you much because you know that the long term trend is still in your favour. If, on the other hand, you had purchased a six months futures contract you would have had numerous margin calls and, finally, you would have been forced to take a loss.

Because timing is so very important, I recommend you study the art of charting and follow the market on a day-to-day basis. An intimate knowledge of moving averages, support and resistance levels, open interest and volume figures can improve your timing tremendously and save you thousands of dollars.

Playing the futures market also requires nerves of steel. A number of excellent books exist on the subject and, inevitably, they conclude that a keen understanding of technical indicators and a lot of discipline are essential. One of the hardest things to learn is to cut losses early and decisively and to let your profits run when prices go in your favour. If you speculate, you will also have to learn to look at the market each morning in order to ask yourself whether the situation is still the same as it was the day before and whether your position still makes sense. The futures market can make you a lot of money and it can lose you a fortune. It will be your ally if you treat it with respect and caution, and it can turn into your worst enemy if you ever feel too sure of yourself.

OPTIONS

Call and put options have existed for gold, silver, platinum and palladium for several years but, unlike options on shares, they were never traded on the exchanges. The leading options dealers were Valeurs White Weld of Geneva, Switzerland, and Mocatta Metals Corporation of New York. In 1979, the Winnipeg Commodities Exchange announced an options program on gold but, unlike the vehicles available from Mocatta or Valeurs White Weld, Winnipeg options were for futures contracts, not for actual physical gold. Shortly thereafter, the first exchange-traded puts and calls for physical gold were introduced by the European Options Exchange in Amsterdam. Several banks in European countries were appointed as depositories for the gold while the actual options contracts were made available to the public through brokers.

In 1982 the concept caught on in North America. A Canadian dealer, Guardian Trust Company, was appointed as the depository for North America and the Montreal and Vancouver exchanges provided the trading mechanism. This means that you can now buy and sell Canadian call and put options on physical gold bullion simply by telephoning your broker.

The market is still in its early stages and, at times, experiences liquidity problems. As a result, quotations are not always very favourable to the investor, but indications are that this will soon change. Not only is the European Options Exchange trying to link up with one of the Far Eastern exchanges so that a 24-hour market can be provided, but several U.S. exchanges have also joined the club.

The first of these is COMEX in New York which recently started an options program for gold futures. Other vehicles are expected to come on stream soon and will include options for physical bullion in much the same way as they are now offered through Montreal and Vancouver exchanges. Professional traders are anxiously awaiting these developments because they will benefit from the minute differentials between various markets which will allow them to arbitrage. And when the traders arrive they will bring with them better volume and liquidity.

Gold, by the way, will not be the only options traded commodity. The Toronto Stock Exchange recently launched a silver options program and most experts think that platinum and palladium will not be far behind.

Advantages:
- You can profit when prices go down.
- Leverage: your money tie-up is substantially reduced.
- Clearly defined risk.
- You invest only in precious metal.
- Options prices are widely quoted.
- No storage risk.

Disadvantages:
- Liquidity is often poor.
- You are tied to certain contract sizes, maturity dates and trading hours.
- Expensive.

Opportunities and Pitfalls:
- Know what your option represents: physical gold or a futures contract.
- Few people understand the options market. If you don't, make sure your broker does.
- Study the literature the exchanges provide.
- Start small, build up as your expertise grows.

How Options Work

As we saw in the last chapter, futures contracts provide you with *leverage*, the ability to play with more money than you actually put up for investment. We also saw that the major drawback with a futures contract is that it represents an open ended risk.

This is where the advantage of an option comes in. With a futures contract you actually purchase or sell a commodity at a fixed price for future delivery. With an option, on the other hand, you only buy *the right* to purchase or sell the same commodity, which means that if the market goes against you, you just write off your initial investment and forget about it. In other words, you have *no obligation* to go through with the deal. The net result is that you can effectively use leverage without having to live with an unlimited downside risk. In fact, you can calculate your risk to the penny because it can never be greater than the option's price.

Understanding options is not a simple matter but the chart below goes a long way in explaining the basics. It was made available to me by the Montreal Exchange whose publication, "Understanding Gold Options", I highly recommend.

Expectations and Motives				
Hedger —wishes to protect against Speculator —wishes to profit from	An increase in the gold price		A decrease in the gold price	
Basic Goal	Lock in a low gold purchase price		Lock in a high gold sales price	
Basic Gold Options Strategies	Buy calls	Write puts	Buy puts	Write calls
This Confers	The right to buy gold at a fixed price	The obligation to buy gold at a fixed price	The right to sell gold at a fixed price	The obligation to sell gold at a fixed price
And Involves	Paying a premium	Receiving a premium	Paying a premium	Receiving a premium

Basic Gold Options Strategies

As you will note, there are two basic types of options: calls and puts. When you buy a call option, you purchase the right (but not the obligation) to buy gold at a fixed price in the future. In return for this right, you pay a price, or "premium", and this price is the entire extent of your risk. Conversely, when you buy a put option, you acquire the right (but not the obligation) to sell gold at a fixed price in the future. Again, you pay a premium.

Writing an option involves an obligation, not just a right. But, in return for undertaking this obligation, you receive a premium. If you write a call you give someone else the right to purchase your gold at a fixed price. If he exercises that right, you have to deliver. Therefore, you have to give evidence to the Exchange that you have the gold at the time you enter this transaction. Writing a call does not cost you a thing, but generates income. The premium you receive comes from the person who bought the right to buy gold from you.

You can also write puts, meaning that you give someone else the right to sell gold at a fixed price to you in the future. In other words, you have the obligation to buy gold if the purchaser of the put exercises his right to sell, and therefore you have to evidence the fact that you have sufficient money. Again, writing a put costs you nothing and you actually pick up a premium.

As our table shows, buying calls and writing puts are strategies to exploit an increase in the gold price, while buying puts and writing calls are maneuvers which work when the price falls. Which one you choose will depend on your particular objectives and on the condition of the market.

Once you understand the basics of options trading and are ready to make a transaction, make sure you think it all the way through. Many investors totally overlook factors such as brokerage commissions or the cost of carrying margin deposits, where necessary. Don't forget to take these into account.

You should also consider whether the newly created COMEX options on futures can be of use to you. Options can eliminate the largest drawback of a futures contract, namely the open ended risk. For the price of the premium, you can buy far more effective protection than a stoploss ever gives — and you still have the advantage of leverage! Moreover, if you place a futures contract and a protective option at the same time, the exchange will recognize that as one "spread", rather than as two independent positions. This means that as long as the option appreciates by as much as the futures contract loses in value, no margin deposits will be called. Obviously this can translate into real savings while eliminating the inconvenience of frequent margin calls from your broker.

But options on gold futures contracts are every bit as complicated as options in bullion — study them carefully before you act.

Finally, I should point out that there are many more strategies available than the four we have explored. Options are highly creative vehicles which lend themselves to a great variety of applications. If they interest you, try to find a broker who is an expert in options *and* has a basic understanding of the precious metals market. This sounds logical, but it may not be easy.

Let me say two more things about options. While they seem to cater to just about every objective, you should remember that they are short term investment vehicles best suited for trading. If you do not have the time to watch them constantly and if you do not want to go through an exercise, roll over or close out every few weeks or months, they may not be right for you. The second point is that options are still very new. The options contracts available from private dealers such as Valeurs White Weld and Mocatta Metals, as well as the exchange-traded options on the Montreal Stock Exchange and on COMEX have all had considerable liquidity problems.

It is my firm belief that options will be among the most successful and accepted investment vehicles a few years down the road, but right now they have the opposite problem: not enough people participate to make their purchase or sale as smooth as bullion, certificates, stocks or futures.

MINING SHARES

The first thing you have to realize when investing in mining shares is that you are not investing in precious metals alone. To an equal degree you are also investing in the future of a corporation, which makes the issue much more complicated. Instead of analysing only the outlook for the precious metal — which is difficult enough in itself — you also have to consider the corporate health of the specific mine you will be purchasing a share of.

Initially, you must familiarize yourself with the background of the corporation, its recent development and its prospects for future production. You should also look at an earnings projection, the capitalization, the dividends, and other important statistical information. In addition, there are a number of factors which pertain specifically to the stock of an operating mine.

The first thing to take into account is how much it costs the mine to produce its precious metal. The production cost usually depends on the grade of the ore mined. In general, the lower the grade, the higher the production cost. Production cost and grade are therefore important investment considerations. You would hardly want to invest in a gold mine if the production cost were $480 per ounce, while the market price was only $375. On the other hand, if the gold price reached $500, a mine with that production cost would have an excellent outlook.

Many smaller mines and mining properties, which are not now operative, may suddenly become profitable if precious metals prices continue to rise. Some of them will be run by their current owners and others will become candidates for a takeover by larger corporations. There are a number of factors which affect the production cost. To name only some of them, there are labour costs, corporate management, taxation changes, and the state of the equipment to worry about.

Advantages:
- You earn dividend income.
- Highly liquid.
- Prices are widely quoted.
- Eligible for tax shelters such as RRSP's, IRA's, KEOGH's.

Disadvantages:
- You invest in more than just precious metals.
- More knowledge required.
- You are tied to "lots" (e.g. 100 shares) if you want to avoid high commission charges.
- Money tie-up.

Opportunities and Pitfalls:
- If you don't understand both the mining business and precious metals, be sure that your broker does.
- When dealing in foreign mining shares, choose a broker who specializes in them.
- Check taxation on foreign shares.

Another consideration when analysing a particular mining property is its "life". This is the length of time a mine can be expected to operate at unchanged production levels. In other words, the expected reserves of gold in the ground determine the mine's life. A mining operation with a very short life is obviously not suitable as a long term investment.

You should also bear in mind that the purchase and sale of mining shares, like other stock market transactions, are subject to a broker's commission, usually somewhere between one and a half percent and five percent. On the other hand, since the investor is normally paid dividends, these should more than compensate.

As you can see, investing in mining stocks is not a simple proposition. The guidance of a good broker is absolutely essential unless you are experienced in mining share analysis, and are also familiar with the price outlook for precious metals. Relatively few sales representatives of brokerage firms are really knowledgeable when it comes to gold, silver, platinum and palladium. If you need help, you should deal with one of the institutions which maintain a high profile in this field.

Canadian Mining Shares

Canada is the free world's second largest producer of precious metals, with about five percent of its gold, platinum and palladium, and about fifteen percent of its silver. Since Canada is also a stable country politically, its mining shares are highly prized even beyond its own national borders.

Most popular with the international investment community are the shares of producing Canadian gold mines, some of which are compared in our table below:

Mine	Outstanding Shares (millions)	Average Grade (oz. gold per ton)	1983E Production (oz.)	Life of Mine	Approximate Production Costs per Ounce
Agnico Eagle	14.1	0.18	49,000	medium	$205
Camflo	4.1	0.12	50,000	medium	$225
Campbell Red Lake	48.0	0.62	220,000	long	$105
Dickenson Mines	12.0	0.31	51,900	long	$250
Dome Mines	70.0	0.26	418,000	short-medium	$205
Giant Yellowknife	4.3	0.23	62,500	medium	$320
Kiena Gold Mines	5.9	0.20	61,200	long	$240
Lac Minerals	24.4	0.18	257,100	medium	$240
Pamour Porcupine	7.0	0.09	102,000	short-medium	$350
Sigma Mines	8.1	0.13	61,500	medium	$190

Selected major Canadian gold mining shares

This table, by the way, was made up from data prepared by the brokerage industry, as well as statistics contained in the annual reports of the various mining companies. Another favourite source of mine is the Canadian Mines Handbook, which is published annually by The Northern Miner (Northern Miner Press Limited, 7 Labatt Avenue, Toronto, M5A 3P2). Although the handbook costs around $20, it contains a wide variety of production figures, financial results and the share prices of over 1,000 mining companies.

Once you know the cost of production and the mine's estimated life, the next question is, what would happen to your earnings if the gold price were to change by $100?

Actually, the impact of such a price movement is quite easily calculated. All you have to do is multiply the mine's production by the dollar amount you expect the gold price to change. If, for example, a company produced 50,000 ounces of gold, and if the gold price suddenly jumped by $100, this would translate into $5,000,000 in added revenue. You can then look up the number of shares the company has outstanding and, by dividing the extra revenue into this figure, arrive at the additional earnings per share.

Of course, your calculation would be somewhat simplified. To begin with, it assumes that the mine's production rate and the grade of ore exploited would stay constant. Secondly, the figures you arrive at are gross figures which would be brought down by operating expenses and taxation.

Obviously, professional mining share analysts have to concern themselves with even more details than these. Such factors include yield, price/earnings ratio, working capital, debt/capital ratio, book value, etc., etc. On the production side, an analyst will try to determine whether a mine's equipment is in good repair and what its exact milling capacity is. During times of depressed precious metals prices, he will try to find out how much metal a company pre-sold in the futures market and at what price. These are vital questions which, after looking at the Canadian Mines Handbook, the various companies' annual reports, and talking to your broker, you should be able to answer before making an investment decision.

The same factors which apply to gold shares also affect the price of silver mining shares. There is, however, one substantial difference. The mines listed above are primarily in the business of mining gold, while other metals are only produced as a by-product. With silver it is the other way around. Most of the white metal is produced as a by-product of copper, nickel, zinc or lead. Thus, it is very difficult to invest directly in the price of silver through the purchase of a major Canadian mining share.

The second half of 1982 illustrates this problem perfectly. Assume that, in June of 1982, you purchased shares in Inco Limited, a leading Canadian silver producer. At the time of purchase, the price of silver was very depressed, somewhere between $5 and $6. Three months later, however, the silver price shot up to $10. What would have been the value of your shares? Well, the fact is, Inco shares actually went down, because Inco's major product is nickel and nickel prices were still falling.

The same, by the way, is true of platinum and palladium, where again Inco is Canada's largest producer. Unfortunately, the firm's platinum and palladium output contributes relatively little money to overall earnings and it is nickel that determines production policy.

Canadian Juniors

A second group of precious metals mining stocks are those in the "junior" category. This term is used to describe mining operations which don't have an established production record and which usually don't pay dividends. Some of these firms are still trying to build up their operations, others own ore bodies which could do exceedingly well under the right price conditions. Junior mining stocks are always purchased for their capital gains potential and are a more risky investment. They are often listed on secondary exchanges, such as the Montreal and Vancouver exchanges.

Advantages:
— Greater profit potential if project succeeds.

Disadvantages:
— High risk: relatively few juniors advance to the senior category.
— Low negotiability.
— Money tie-up.
— Expensive (transaction costs are unusually high).
— Not a direct investment in precious metals.
— Price quotations more difficult to obtain.

Opportunities and Pitfalls:
— Avoid companies with debt.
— Make sure the shares you purchase can be traded.
— If you invest several thousand dollars, diversify.
— Do not purchase penny stocks unless you are prepared to lose every penny.

Many investors hold a very small percentage of their money in junior mining shares because of their enormous appreciation potential in an environment of rising gold prices. This, however, can also work in reverse. Many junior stocks which traded at $5 in early 1980 were worth only a few cents a few years later.

One problem with junior mining shares is that you often lose in liquidity what you gain on paper. Make sure that the shares you are interested in are actually listed and check on how negotiable they are.

Another group of shares which may interest you are those of exploration companies. These firms don't even produce gold. Instead, they own properties which increase in value as the bullion price rises. Once their price objective is met, the property is sold to a larger operating mine for exploitation.

As is the case with junior mines, shares of exploration companies tend to outpace senior stocks in a positive price environment. But be careful: if the gold price doesn't move in the right direction, these shares are very vulnerable on the downside.

The shares of major producers, junior shares, as well as shares of exploration companies all have one great advantage. If you are a Canadian resident, you can deposit them into Retirement Savings Plans, provided they are listed on a Canadian exchange. This allows you to defer capital gains and income earned on your shares into the future. Americans are even more fortunate: tax deferral plans such as KEOGH's and IRA's also accept foreign share-holdings. If you are a U.S. resident, the only drawback is that your dividend income from Canadian sources will be reduced by a withholding tax of 15%.

South African Mining Shares

South Africa is the world's largest gold producing nation and, as such, features a whole array of gold mining shares. Almost all of these represent a very direct play in gold and have an excellent production and dividend record. Dividends, in fact, are easily their most popular feature and you will quickly see why when you look at our table on page 112.

Another advantage of South African mining properties is that their ore reserves are very sizeable, especially when compared to North American mines. At least ten of the major gold producing companies pride themselves on ore bodies with a life of more than twenty years! My favourite South African mines, by the way, are Driefontein Consolidated, Kloof, Southvaal and Vaal Reefs. All four represent conservative long-term investments which will generate attractive dividend income in the meantime.

South African shares are actively traded on the Johannesburg and

Stock	Outstanding Shares (millions)	Average Grade (oz. gold per ton ore)	1983 Production (oz.)	Life of Mine (yrs.)	Approximate Production (cost per oz.)	1983 Dividend (%)
Blyvoor	24.0	0.2411	592,000	Short-Medium	$220	10.8
Bracken	14.0	0.1190	115,800	Short	$290	19.4
Buffelsfontein	11.0	0.2958	971,830	Long	$220	11.5
Doornfontein	10.0	0.2186	320,060	Long	$255	6.2
Driefontein Cons	102.0	0.4195	2,391,530	Long	$120	8.8
Free State Geduld	10.4	0.2186	873,310	Long	$295	6.9
Harmony	26.9	0.1318	1,025,810	Long	$320	9.2
Hartebeestfontein	11.2	0.3200	972,610	Long	$190	8.3
Kinross	18.0	0.2009	392,920	Medium-Long	$200	6.1
Kloof	30.2	0.4887	1,001,850	Medium-Long	$120	5.8
Libanon	7.9	0.1921	322,730	Medium-Long	$235	6.9
President Brandt	14.0	0.2193	772,220	Long	$220	7.9
President Steyn	14.6	0.2120	833,720	Long	$230	7.8
Randfontein	6.1	0.1608	952,950	Medium	$160	7.2
St. Helena	9.6	0.1967	446,480	Long	$210	7.4
Southvaal	26.0	0.3633	1,174,560	Long	$130	5.0
Stilfontein	13.1	0.2218	396,870	Short	$290	11.8
Vaal Reefs	19.0	0.2251	1,396,240	Long	$170	8.0
Western Deep Levels	25.6	0.3600	1,268,230	Long	$165	5.4
Western Holdings	14.3	0.1419	1,292,850	Long	$270	9.7
Winkelhaak	12.2	0.2015	472,780	Long	$160	7.9

Selected major South African gold mining shares.

London exchanges and, over the counter, in the United States. Because Canadian or U.S. holders are considered non-residents, there is a fifteen percent withholding tax which is deducted from dividends. You should also know that not every broker can assist you in trading these issues, although the list is steadily growing.

A number of major South African gold shares are available in the U.S. as "ADR's", or American Depositary Receipts. An ADR is really a storage receipt issued by a U.S. bank for shares it holds in a safekeeping facility abroad. Although you shouldn't make your share selection dependent on this point, it does help if it is available as an ADR. Its negotiability, its registration and its dividend flow are a lot more efficiently organized than with physical holdings of a foreign share.

Be careful, however, when you buy ADR's. They are not always quoted on a par with the actual shares. Quite often, one American Depositary Receipt represents ten, or even a hundred shares. So don't rush to your broker asking him to buy 500 ADR's of a particular issue if what you really want is 500 shares. If you do, you may end up with 5,000 or 50,000 shares!

Even easier than owning ADR's is to buy a foreign stock which is actually listed on a North American stock exchange. Most South African mining shares, unfortunately, are not.

Evaluating a South African mining share requires the same knowledge as any other stock. If you are a conservative investor, you should concentrate on mines with a long life and a low cost, because these can be expected to increase their dividend as the gold price rises. Moreover, since there are so many excellent mining properties to choose from, you should diversify your portfolio as much as you can. This has two advantages. It protects you against bad management in any one mine and, at the same time, it provides you with a safeguard against sabotage.

While the chances of sabotage are not very great, you should seriously think through how it would affect you as an investor. Let us assume that your favourite gold mine became the victim of a terrorist attack and several of its major shafts collapsed, thus bringing production to a standstill. The inevitable result of such news would be a sharp rise in the gold price and, along with it, the plunge in value of your shares.

Perhaps the best way to protect yourself is by buying shares in one of the many diversified South African investment houses. Companies like Anglo-American Corporation, De Beers Consolidated Mines Limited and Gold Fields of South Africa Limited have holdings in a great number of operating mines and, as a general rule, fluctuate up and down with the value of precious metals. One problem with these shares, however, is that they represent an investment in more than just gold. Most also have holdings in coal, diamonds and even oil, although their interests in gold are normally high enough that the bullion price determines their value.

By far the most popular of these shares is ASA Limited, formerly American South African Investment Company Limited. Listed on the New York Stock Exchange, ASA provides investors with asset diversification in the minerals and metals sector. Here is a listing which summarizes ASA's holdings as of February 2, 1983:

INVESTMENTS HELD BY ASA LTD. IN EARLY 1984
(Number of shares)

Long life, high grade gold mines

Driefontein Cons.	3,158,000
Kloof	1,257,000
Western Deeps	150,000
Southvaal	1,273,000
Vaal Reefs	720,000
Unisel	284,000
Kinross	215,300
Winkelhaak	1,216,700

Major gold-uranium mines

Hartebeestfontein	242,600
Buffels	483,000
Zandpan	988,300
President Steyn	503,100
St. Helena	432,100

Miscellaneous Holdings

Anglo American (finance house)	121,000
Amcoal (coal)	553,200
De Beers (finance house/diamonds)	1,000,000
Impala (platinum)	268,700
Palabora	150,600
Rustenburg (platinum)	348,712
Sasol (oil/coal)	1,875,000
TCL (coal)	582,300
Trans Natal Coal (coal)	810,000
Samancor (manganese)	521,000

As you can see, ASA gives you access to most of the long life and low cost gold mines, as well as to other holding companies and direct interests in uranium, coal, diamond, oil and platinum group metals. Interestingly, when the gold price rebounded in mid-1982, ASA appreciated at an even faster rate.

One minor problem with South African shares is that their profits and dividends are originated in South African rands. Thus, the rand/dollar exchange rate has to be a factor in your decision. At present, however, the rand is very depressed, which means that now is the time to buy. As world prices for metals and minerals start to recover, the rand's value should go up along with them, making the dividends you will get even more attractive.

But what about other precious metals? Isn't South Africa also the world's largest producer of platinum, and the second largest producer of palladium? Luckily, South Africa does have two mines whose major earnings are derived from platinum group metals. The world's two largest producers, Impala Platinum Holdings and Rustenburg Platinum Holdings are listed in Johannesburg, London and, again, over the counter, in New York. Even during difficult periods dividends often remain at the high levels typical for South African precious metals shares. Because the management is aware that a high dividend is one of the prime attractions in foreign markets, sharp declines in earnings are usually not fully reflected in the dividend. But bear in mind that the same is true in reverse. South African mines tend not to distribute earnings fully during times of very strong prices.

You should note that Impala and Rustenburg are the only two investments which give you a relatively direct interest in platinum and palladium and which are traded internationally and very liquid. If you are convinced by the case for these two metals, (and more will be said about them in our section on strategic metals), you should seriously consider their purchase.

U.S. Mining Shares

The United States still contributes heavily to the world's gold and silver output. Exciting new ore-bodies are being mined but, the trouble is, most major U.S. mining firms have grown and diversified to such an extent that they no longer represent a direct play in any one metal. The Homestake, for example, is still America's largest gold mine, producing about 300,000 ounces annually. But it is now equally interested in its uranium, lead, zinc, silver and forest product operations.

Of course, diversification can be an asset. As a rule, it lends a share more price stability and makes your investment less vulnerable to drastic price declines in just one metal. But, on the other hand, as you have seen in our example with Inco in Canada, a rise in precious metals can go ignored when they are not a major profit source for the company. In other words, if you want to invest in gold, as opposed to mining in general, buying shares in a conglomerate is not the answer.

One exception may be Newmont Mining Corporation. This is a diversified resource firm with a relatively high interest in gold, although right now most of its capacity is devoted to copper. Still, gold mining is taking on a more and more significant role — so much so, in fact, that Newmont may soon become the leading U.S. gold producer.

THE 15 LARGEST U.S. SILVER PRODUCERS

Mine	Average Annual Production (millions of oz.)	Impact of Silver Price on Share Performance
Hecla Mining (includes Day Mines)	5.5	1
Asarco Inc.	4.0	2
Anaconda Co.	3.4	3
Kennecott Corp.	3.2	3
Sunshine Mining Co.	2.4	1
Phelps Dodge Corp.	2.2	3
Gulf Resources & Chemical Corp.	1.9	3
Callahan Mining Corp.	1.9	1
Homestake Mining Co.	1.8	2
Occidental Minerals Corp.	1.7	3
Newmont Mining Corp.	1.3	2
Amax Inc.	1.3	3
Duval Corp. (Pennzoil)	1.2	3
Coeur D'Alene Mines Corp.	1.0	1
Cyprus Mines Corp.	0.8	2

Impact of silver price fluctuations on share price
    1 = relatively pure silver play; impact high
    2 = reasonable diversification; reduced impact
    3 = high degree of diversification; impact negligible

Most U.S. shares which do offer direct metals plays, by the way, are in the silver sector, an important fact given the lack of relatively pure

South African and Canadian silver participations. The most popular among the major American silver producing mines are Bunker Hill, Callahan Mining, Coeur d'Alene, Day Mines, Golconda, Hecla, Silver King Mines and Sunshine Mining. In addition, there are dozens of junior mining properties which your broker should be able to identify.

Be careful, however, because U.S. junior stocks have the same advantages and drawbacks as Canadian issues. They can be put into your IRA or KEOGH, but they are often traded on smaller exchanges like Spokane or Denver, where liquidity can be low and commissions high. To Canadians, of course, they are not eligible for an R.R.S.P. and are subject to the usual 15% withholding tax.

PRECIOUS METALS FUNDS

A variety of precious metals funds have recently become available to the North American public. As a general rule, the investor can purchase individual units or shares which then go up and down according to the fund manager's performance. You pay for this service in the form of a fee which is usually charged to the fund and often amounts to a very modest amount of money.

Precious metals funds fall into two major categories: *mutual fund trusts* and *investment corporations*. Most mutual funds are adjusted to their net asset value on a regular basis. In other words, the managers calculate the total value of all the fund's holdings once a week or month. The purpose is to establish a value basis, at which new investors are allowed to buy units, or at which existing unit-holders can sell. This is an important point, because it guarantees that the value of your investment fluctuates in line with the value of the assets held by the fund. If the market goes down, so do your units — if it advances, you can be sure that you're on the up side as well. And when you want to take your profits, you can depend on getting your percentage of the fund's net worth.

One disadvantage to many mutual funds is that they carry heavy commission charges when purchased from a broker: usually between five and ten percent! An exception to this are "no load" funds which are sold to the public on a direct basis and therefore carry no commissions.

The second broad category of precious metals funds is that of investment corporations. Very few of these exist in the United States, but in Canada they are so popular that no less than five of them were successfully launched in 1983 alone! Unlike mutual funds, investment corporations have a fixed operating capital which is usually raised in a public underwriting. Once the corporation is funded, its shares freely trade at the exchanges, depending on what the market thinks their value should be. Unfortunately, the marketplace has a tendency to overlook certain things, and this can cost you dearly. In 1983, for instance, I helped raise $50 million for BGR Precious Metals Inc., an investment corporation in which I share management responsibilities with N.M. Rothschild, a leading London bullion dealer. Although we quickly increased the net asset value per share by over 10%, in the open market our shares were soon selling at a handsome 20% discount! Why? Because the public saw the bullion price decline and deduced that BGR would automatically fall in tandem.

This simple example shows the risks — and the opportunities — of investing in one of the new investment corporations. At certain times, the shares will sell at a discount, frustrating you if you would like to sell. To an astute buyer, on the other hand, they may represent the only opportunity to buy bullion and shares at a hefty discount!

The safest route is to contact your broker or the funds manager to obtain a prospectus before you commit yourself to the purchase of units in any particular fund. Pay special attention to the section dealing with selling commissions and management fees. You should also examine the performance of the fund and familiarize yourself with the strategy and objectives the managers use in running it. Some funds have the policy of always being fully invested in the market, no matter whether it is in a down trend or an up trend. This may be exactly what you want and then again, it may not. Finally, make sure that the fund units you are considering for purchase are traded with some liquidity, so that you can sell them again if and when you want to.

*—Note: The author also manages the bullion portion of the **Dynamic-Guardian Gold Fund,** a no-load fund investing in physical precious metals and mining shares. (See "Dynamic Funds Management Limited", page 169.)*

MANAGED ACCOUNTS

Managed accounts, like precious metals funds are only as good as the experts who run them. However, there is one important distinction:

while a unit fund has to appeal to many investors and can only have a very general strategy, a successful account manager can tailor his approach to the market with your objectives in mind.

Account management is usually available from investment firms, brokerage houses and bullion dealers, and generally requires a financial commitment of not less than $100,000. A great number of funds managers compete for this business in the offshore havens and, to a lesser degree, in the United States. In Canada, there are very few institutions which engage in the managed accounts business.

Advantages:
— Convenience.
— Professional management.

Disadvantages:
— Management fees.
— Risk of poor performance.
— Money tie-up.
— Liquidity usually poor.

Opportunities and Pitfalls:
— Examine manager's track record.
— Meet account manager and make sure you agree.
— Start with the smallest permissible amount.
— Negotiate fees if a large amount is involved.
— Check cost of liquidating account.

If your financial holdings are significant, a managed investment program can be an excellent option. Particularly if you are a professional in your own field and have neither the time nor the expertise to follow developments in the precious metals market, a managed account may be just what the doctor ordered. Most firms devote considerable time to such accounts because it is in their interest to impress their client and get a larger share of his business. At the same time, management fees are usually quite reasonable. You should realize, however, that a managed investment program is discretionary. In other words, you have absolutely no control over the manager's investment decisions, and you generally have to give him full powers to use your money as he sees fit. In addition, you have to agree not to hold the manager responsible if he happens to make a mistake and you lose your money.

The best safeguard is to get to know a manager very well before you entrust him with your account. Try to find out what his track record is and don't be shy about asking him to document it. Having friends or acquaintances who have already had positive experiences with the firm you are considering is a big advantage. Once you are convinced that the manager is an expert in his field and can make you money, explain to him carefully what your investment philosophy, your financial standing and your expectations are, and make sure that he understands it. You should also examine the balance sheet of the firm he represents and, if possible, obtain references.

Once you have done all this, and are happy with the results, place as small an amount as the firm will take and give the manager a fair chance. Try to spend time with him every three months to review his performance. This will help you understand his style and strategy and, in turn, he will get to know you better and do a better job. After a year or so, you will know your manager reasonably well and, if all goes the way it should, you will trust him. That is when you can deposit additional funds.

Good money managers are extremely rare, and the best are not in the business of running $100,000 accounts. Unless you are not happy with his performance, try not to phone him every few days to see "how things are going". Remember that you want your manager to be your ally and not someone who sighs when he hears your voice. You can rest assured that your manager is every bit as aware of the money you have given him as you are.

COLLECTIBLES

The only area of precious metals investments we have not yet explored is that of collectibles, a term which I should further define. All the vehicles we have so far discussed have one thing in common: they are purchased only because they represent an investment. Collectibles, on the other hand, are items which are also purchased for their beauty, historical significance or educational value. In other words, the cool judgement of an investor is here complemented with emotional sentiment.

One of the first things professional traders learn is that investment considerations and emotions don't mix well. Collectibles such as numismatics or jewelry illustrate this point. The fact that they may appeal to you personally does not necessarily mean that they also appeal to other people, which can translate into a tremendous loss of liquidity just when you need it most. Picture yourself wanting to sell your favourite ring in an emergency and try to assess how others would react to such an idea. Quite likely, you would get much less money for it than you paid. The same can be true of numismatic coins or other collectibles made out of precious metals. It is therefore essential that you don't tie up too much of your money in them and *don't consider them as an investment*.

Advantages:
— Esthetic appeal, historical significance, etc.
— Potential as a hobby, not "just an investment".

Disadvantages:
— High cost.
— Low liquidity.
— Not a direct investment in precious metals.
— Adverse taxation (sales taxes in most states and provinces).
— Emotional appeal may cloud your judgement.

Opportunities and Pitfalls:
— Don't buy collectibles as investments — buy them because you like them!

Numismatic Coins

In contrast to modern bullion coins which are minted in enormous quantities and whose price depends strictly on the content of bullion, numismatic coins are governed by four major factors: their rarity, their age, the quantity originally produced and their condition.

A numismatic dealer's inventory may range from items minted in 500 B.C. to the 1982 Constitution coin issued by the Royal Canadian Mint. In recent years, an increasing number of governments have issued coins to commemorate events in their history. Aside from the factors mentioned above, numismatic coins are also collected for their beauty, historical significance, their precious metal content and their educational value.

Numismatic coins do depend on the price of metal to some degree, but it is usually a minor factor. Thus, their price is usually far higher than the value of the gold or silver they contain and their values fluctuate to a much wider extent. For example, a rare $5 gold piece may contain $80 worth of gold and may sell for as much as $800.

The minimum you can recover from a numismatic coin investment is always either its face value or its metal content. But this is not an entirely realistic standard of value, because in order to sell the metal you would have to face assay and melting charges which can be significant.

NUMISMATIC GRADINGS

1. **Proof**
   Coins struck specially in small numbers for investment, presentations, and other purposes. These coins are usually struck twice ("double struck") and as a result have a high, mirror-like finish. Due to the small number produced and the great demand among collectors, proof coins usually sell at a considerably higher price than the standard issue.

2. **Brilliant Uncirculated (B.U. or U & C)**
   These coins are struck on regular dies, although they are not intended for circulation. Most bullion type coins and most numismatic coins produced today remain in brilliant uncirculated condition. In other words, their holders try deliberately not to damage the coins in order not to reduce their investment value.

3. **Extremely Fine (E.S.) or Extra Fine (X.S.)**
   These coins are slightly circulated and show some signs of wear, but still have their original polish.

4. **Very Fine (V.F.)**
   Details and design are still clearly marked; only the highest surfaces are worn down slightly.

5. **Fine (F.)**
   Coins which have been circulated and show definite signs of wear, especially on the finer parts of the image. However, all the details are visible. The grading "fine" is the minimum standard for coin collectors.

6. **Very Good (V.G.)**
   The inscriptions and design are still clear and bold, but worn.

7. **Good (G.)**
   Design and inscriptions are readable but quite worn.

8. **Fair (F.)**
   Features are still identifiable but not easily readable.

9. **Mediocre (M.)**
   Very worn or damaged.

10. **Poor (P.)**
    Unless it is very rare, a coin in "poor" condition usually doesn't fetch more than the value of the metal it contains.

In the numismatic sector, probably more than anywhere else, the selection of a reputable and long established dealer is essential. Numismatics are a complicated field and require a high degree of expertise and knowledge. Coins are sold in various conditions, along a "grading scale" which ranges from *poor* to *proof*. Determining the right grade is very difficult, although the price of each coin is highly dependent on it. With older coins, particularly those from ancient times, evaluation is even more difficult: in many cases only your trusted dealer can separate an authentic ancient coin from a forgery.

The numismatic coins of recent years help you overcome this problem. Almost all of them are encapsulated in plastic and, unless you tamper with them, will therefore retain their original condition. On the other hand, collecting modern coinage is not nearly as interesting and exciting as collecting coins from ancient Greece, Rome or Constantinople.

Medals and Medallions

One area you should stay away from altogether is that of "medallions", unless they are struck by governments or are of historical value. Medallions are not legal tender and therefore never show a face value. Most of you will have received brochures advertising such "coins" made by private mints and refining institutions. Printed and distributed at a very high cost, these offerings are always directed at the "serious investor". In reality, the investor usually pays a hefty premium for all the expenses of the mail order promotion which, even after significant appreciation of the precious metal, often make it impossible to recoup the original investment. Don't expect any dealer to pay a premium for the esthetic appeal of such a private issue. The best you can usually get for such items is the market value of the metal it contains — minus melting and refining charges!

Jewelry

"If things get really rough, I can always trade in my wife's jewelry" is a sentence that best illustrates the widespread belief that jewelry is a safe investment.

Jewelry made of precious metals offers the classic appeal of an incomparable gift, but it cannot be seriously recommended as a good investment. To begin with, precious metals in their pure form are not suitable for use as jewelry. They are usually alloyed to standard finenesses, such as 22, 18, 14, 12 and even 10 karat in the case of gold. This process creates a cost which is passed on by the refiner. Once the craftsman takes delivery of this alloy, he charges for the considerable workmanship and expertise he has put into the jewelry he makes. Then, on the way from his workshop to the store shelf, jewelry is heavily taxed and an array of excise, wholesale and retail mark-ups are added to the base price. Not only is it very unlikely that you will be able to recover these additional costs but, in a crisis, you may even have to pay refining costs when you try to sell.

---

fact:

In 1967, the U.S. Federal Trade Commission defined "solid gold" as "any article that does not have a hollow centre and has a fineness of ten karat or higher"... Investors beware!

In this connection, I am again reminded of the plight of those Vietnamese refugees who came to the U.S. a few years ago. While most brought gold wafers with them, some carried jewelry. Those carrying bullion had to pay assay charges because their wafers were not made by internationally recognized refiners. However, the three or four percent charge was a small price to pay for the convenience of being able to instantly acquire local purchasing power.

Now, here is what happened to those refugees carrying gold jewelry: they received the equivalent of the precious metals value after deduction of melting and assay charges. But they lost everything they had paid for in the production of the jewelry, the taxes applied to it, and the selling mark-ups. In most cases, they recovered no more than one-sixth of their original investment!

BORROWING AGAINST YOUR PRECIOUS METALS

You don't have to read a book on precious and strategic metals to learn about the advantages of borrowing. Having access to credit can increase your purchasing power and, particularly in markets where volatile corrections can occur within a very short time, this can be a tremendous advantage.

Just think of the thousands of investors who tied up their disposable capital in gold, but bought too late or too early. Most of them looked on helplessly as prices declined further and further and then, when prices started to rise again, they were unable to participate becaue all their money was already tied up in the market. Financing could have solved their problem. Unfortunately, borrowing money for the purchase of precious metals is not always as simple as borrowing money for more conventional purposes.

That is why lending programs for precious metals are offered by the majority of bullion dealing banks. Most of them are not "aggressive" lenders: the minimum loan required is often as high as $50,000 and most restrict the size of a loan to a maximum of 50% of the total value of the precious metals which serve as collateral. On the other hand, their charges are usually reasonable, generally two to three percent above prime.

Margin Loans

A margin loan provides you with the ability to put up as security something which you want to buy more of. *Buying gold on margin*, in other words, means that you go to a bank in order to buy more gold than you can actually afford to pay for. In practice, unless you make a profit, you will never get to see your gold or your money. The bank takes the money that it has theoretically lent to you, uses it to buy gold on your behalf, and then keeps the gold it has purchased as a security against your loan.

If the bank were to finance fifty percent, a deposit of $100,000 into a margin account would allow you to hold $200,000 worth of gold. But, remember, if you borrowed to the full extent and indeed purchased $200,000 of gold, you would have to be prepared to stand by with more money in case the market declined. In short, if the value of bullion dropped sufficiently to make your original investment worth only $180,000, the bank would ask you for an additional deposit of $20,000. If you could not come up with it, the bank would simply sell enough gold to bring your margin ratio back into line.

As you can see, a bank's risk is quite limited if its margin ratio is as high as fifty percent. Nevertheless, the larger banks do not announce their activities in this field as a service package but prefer to negotiate a loan with each client individually. This is particularly true of organizations with a vast branch network, where the relationship between a customer and the local manager is often a more important consideration than the risk of lending money against bullion. As a result, it is impossible to compare the terms of organizations such as Bank of Nova Scotia, Citibank or Swiss Bank Corporation against smaller institutions who specialize in this field. Margin lending programs whose terms are known and advertised are offered in the United States by Republic National Bank and by Westcoast Bank. In Canada, Guardian Trust Company offers a similar service.

Republic National Bank's and Guardian Trust's financing packages call for a minimum loan of $50,000 and offer competitive rates. Guardian has the added advantage that it also accepts platinum and palladium as collateral.

Westcoast Bank's program is more expensive, but it allows investors to borrow as little as $10,000. The package is retailed through a number of regional bullion firms who are not themselves in the lending business. Most prominent among these is Manfra Tordella & Brookes, a leading New York dealer.

Collateral Loans

In contrast to a margin loan, a collateral loan does not imply that you want to buy more of the same commodity which you then put up as security. Instead, you may well want to borrow against gold you already have in order to buy more Swiss francs or, for argument's sake, more municipal bonds, because their market is right at a particular time. In other words, many investors actually take advantage of their gold holdings in order to make profits somewhere else at the same time. The bank will again want to hold your bullion for you, but it will also purchase whatever security or investment you had in mind or it may even give you the money to make the purchase yourself.

By using collateral loans, you obviously have all alternatives open to you, which is why I recommend them highly. But it would not be fair to give you this advice without warning you that most banks do not encourage this type of loan. A collateral loan allows you to spend the borrowed money elsewhere while a margin loan forces you to do even more business with the lender.

One final point to remember, both on margin and on collateral loans, is that they are in the category of demand loans. This means that the bank can ask you to pay up at any time, although this privilege is rarely exercised. In short, before taking on such a loan, take a good look at all your assets and try to identify at least one other vehicle that you can liquidate in a real emergency.

WHERE TO STORE PRECIOUS METALS

Most people buy bullion not only because they want to make a profit, but also because they deeply distrust paper money. Thus, holding precious metals in physical form is as much a philosophical choice as it is an investment decision. That is why you have to delve back into the history of money when you consider where you should store your bullion.

Should you keep your hoard at your home, where you have ready access to it, but where it is exposed to the risk of theft? Or is it better to keep it in a safety deposit box at your bank, where it is safe from burglars, but where you cannot get to it whenever you wish?

These concerns are as old as humanity itself. The reason why we relate to precious metals the way we do is because we know them to be indestructible and we regard them as the ultimate in portability and international purchasing power. We may not know what a real economic or political emergency would look like, but we all suspect we would be better off with a little bit of bullion.

Given that history abounds with examples of governments which suspended convertibility between gold and paper, which declared forced bank holidays, and which even outlawed the private ownership of precious metals, is it surprising that all of us have a slightly uneasy feeling when we think of keeping our gold at a bank? Even as a banker, I have to admit that I understand this feeling. But as a realist, I have to say that government confiscation of your gold is not very likely — not nearly as likely, in any event, as someone breaking into your house. But this doesn't mean that you should not be careful and observe a number of important points when you put your bullion in storage.

First and foremost, you should select an institution whose integrity and financial standing are beyond any doubt. If you take a safety deposit box, get one at one of the larger branches of a bank where the cost is just the same but the security is often a lot better. On the whole, the safety deposit box is still the best solution if your investment in bullion is relatively small.

If you deal in larger sizes, things become much more difficult. Your desire to keep part of your wealth private will inevitably increase, but safety deposit boxes will no longer be practical. Not only is their size very restrictive, but you also lack liquidity in that you cannot dispose of holdings and repurchase them simply by a telephone call. Far more suitable for this purpose are safekeeping arrangements which can be entered into with most bullion dealing banks. In other words, your metals are held for a fee, just as they would be in your deposit box, but your ability to "use them" will increase substantially. For example, you can call the bank and ask them to sell precious metals out of your safekeeping account when the price is high, and then do the reverse when the price declines.

Banks safekeep precious metals on an unallocated basis, or on a fully segregated basis, and the difference between the two is vital. Unallocated metals are those mingled with the assets of other clients and those of the bank itself, while segregated metals are clearly defined as yours and are generally defined by bar number on your safekeeping receipt. Because unallocated metals appear on the liability side of the bank's balance sheet, they can be used to pay off its debts in the event of bankruptcy, while segregated metals do not appear in a bank's balance sheet and are indisputably yours. Quite obviously, segregated safekeeping is more expensive than unallocated safekeeping, but the difference in price is well worth it. Most bullion dealing banks will open a fully segregated account for a fee of between 1/4 and 1/2 of a percent of the total value of the safekept assets. There is usually an annual minimum charge to discourage smaller accounts.

The terms "unallocated" and "fully segregated" are used in Canada, Great Britain and the central European countries in the same way as the terms "fungible" and "non-fungible" are used in the United States. Under the concept of fungibility, in other words, your metals would be pooled together with those of other investments, while a non-fungible concept would provide you with full segregation.

But segregation or non-fungibility is not the only thing an investor may want. Increasingly, many investors feel that their precious metals holdings are no one else's business and should be kept totally private. Unfortunately, nothing could be less private than a formal safekeeping arrangement with a bank. In many cases, such accounts cannot even be opened without the registration of your social security number, which means that the government can always find out what you hold in your account. Fortunately, there are alternatives.

Canadians concerned with anonymity can simply buy bullion and ship it abroad, or buy it directly for foreign delivery. Current tax legislation does not require any reporting until a capital gain or loss is realized, or until income is received. U.S. residents are not nearly so lucky. Any cash transaction over $10,000 has to be reported to the IRS, as does any cash or valuables transfer of over $5,000. Moreover, Americans have to report assets held in foreign bank accounts once every year.

But where there's a will, there's a way, particularly in a free market system. In the past few years, a great number of storage facilities have been built in the U.S., and many more will be added in the near future. Because they have to make a living from safekeeping alone, they are usually far better organized and equipped than the bank vaults we are used to. For instance, most firms offer 24 hour service, which allows you access to your box at any time. Another advantage is that their private ownership allows them to by-pass banking regulations. As a result, it is not necessary to identify yourself when you enter into a safekeeping arrangement.

Various safekeeping companies cater to different client objectives. Perpetual Storage Inc., one of the nation's largest facilities, is deliberately located in a valley outside of Salt Lake City. The Security Centre of New Orleans, by contrast, owns the former

Federal Reserve vault right in the centre of the city. Far more modern are the facilities of Swiss Security Systems in Miami and Palm Beach. Building and security standards reflect the latest technology, and the flexiility available to the investor is excellent. Another excellent facility which has just been completed is that of Guardian Safe Deposit in Arlington, Virginia. Located minutes from downtown Washington, it features 24 hour armed guard services and numbered accounts facilities sufficient to attract a large number of investors, as well as less expensive bulk space designed for professionals who want to store computer tapes, discs or documents.

But convenience and privacy are not the only things to consider. Make sure you use the same standards of evaluation that you would before entrusting your bullion to a bank. In other words, examine the financial integrity of the institution you are dealing with and don't be shy about inquiring into the details which govern the insurance of your assets.

Finally, compare the cost differences between a bank facility and a privately owned vault. You will find that in most cases private firms are considerably more expensive, but then you do get a lot more for your money. In the final analysis, your decision should be determined by how strongly you feel about privacy and unlimited access to your precious metals.

SELECTING A DEALER

The modern gold market provides investors with a great variety of risk factors. As you have seen, the possible investments range from straight bullion, for which you have to pay in cash, to highly leveraged futures contracts and options which allow you to buy far larger quantities than perhaps you can really afford.

Once you have decided what risk factor suits you, and what investment vehicle is the most adequate for your requirements, you should select a dealer. Finding the right financial institution, dealership or brokerage house is not easy. deally, you want a firm which is long established, can give you professional guidance, and offers a variety of investment vehicles. The table on page 131 gives you an overview of the major retailers offering precious metals services in the United States and in Canada. Their names and addresses, along with those of many other firms whose services I have referred to, also appear in the dealer listings on pages 167 to 174.

Remember that commissions, storage charges, management fees and other investment costs will vary from dealer to dealer. These charges should be carefully compared but, obviously, they cannot be the only decisive factor. Don't expect bullion dealers to engage in

Margin and Collateral Loans	Segregated Safekeeping	Telephone Trading	Numismatic Coins	Certificates or Deposit Receipts: Gold	Silver	Platinum	Palladium	Bullion: Gold	Silver	Platinum	Palladium		
•		•	•					•	•	•		A-Mark Precious Metals Los Angeles	MAJOR U.S. RETAIL DEALERS
•	•			•	•			•	•			Bank of Delaware Delaware	
•	•			•	•							Citibank, N.A. New York	
		•	•	•	•	•	•	•	•	•	•	Deak-Perera major cities	
•		•		•	•			•	•			Dreyfus Gold Deposits New York	
•	•			•	•			•	•			First National Bank of Chicago Chicago	
•		•	•					•	•	•		Manfra Tordella & Brookes New York	
•	•	•	•					•	•			Monex International Newport Beach, CA	
•	•			•	•			•	•			Rhode Island Hospital Trust Rhode Island	
•	•			•	•			•	•	•		Republic National Bank of New York New York	
•	•			•	•			•	•			Bank of Nova Scotia nationwide	MAJOR CANADIAN RETAIL DEALERS
•	•			•	•			•	•			Canadian Imperial Bank of Commerce nationwide	
•	•	•		•	•	•	•	•	•	•	•	Guardian Trust Company major cities	
		•	•	•	•			•	•	•		Deak Perera (Canada) major cities	

long conversations with you on the telephone. During banking hours, they are usually battling with a barrage of incoming calls. However, they will generally be happy to mail you some background material on their services which will help you make the right decision. Dealers are also not in the business of giving market opinions. If you are not well known to an institution, don't try to call their precious metals dealers and ask them for trading advice. They simply do not have the time and it is not their function.

If you are interested in getting professional opinions, you should subscribe to a newsletter or forecasting service. (See the listings at the end of this book.) If your investments are more sizeable, it may even be worth your while to retain a recognized expert as your consultant.

HOW MUCH TO INVEST?

Deciding the percentage of your total assets which you want to invest in precious metals is crucial. It will depend on your age, your social and financial obligations, and your personal assessment of the seriousness of current economic and political problems.

Someone who is young can afford to take larger risks and can therefore invest a larger percentage than others. Someone who depends on retirement income, on the other hand, will have to be more cautious. A person who is married with children has certain obligations to meet and has a responsibility towards his dependents. Someone who is single can afford to be more aggressive in the market.

If you are convinced that ever higher deficits will lead to a new round of inflation and, eventually, the collapse of our monetary system, you will want a maximum of protection. But if you regard inflation as merely a periodic occurrence that eats away at your savings, your precious metals investment should be smaller.

Because of gold's portability and its universal liquidity, I personally believe that the yellow metal should make up the larger portion of your commitment in precious metals although, as we have seen, more money may eventually be made in silver and palladium.

The table shows you approximately what percentage of your financial worth should be invested in gold. These figures only take into account what you need to protect yourself against economic or political uncertainty and do not include gold bought for trading or speculative purposes.

While I think silver, platinum and palladium are excellent long term investments, I do not recommend them for insurance purposes.

HOW MUCH GOLD YOU SHOULD OWN

Net Worth / Age Group	Up to $50,000	$50,000-$100,000	$100,000 $250,000	$250,000 $500,000	$500,000 $1,000,000	Over $1 million
20-30	10-20%	12½-22½%	15-25%	15-25%	15-30%	17½-30%
30-40	10-17½%	10-17½%	10-20%	10-20%	15-25%	17½-30%
40-50	7½-15%	10-15%	10-15%	10-20%	10-20%	12½-25%
50-60	7½-12½%	7½-12½%	7½-15%	7½-15%	10-15%	10-20%
over 60	5-10%	5-10%	5-12½%	5-12½%	7½-15%	7½-17½%

To determine Net Worth add up all your assets and deduct all your debts.
If you are single and have no dependents you can take greater risks: add 5% to proposed gold holdings.
If you have a family or have dependents you should avoid excessive risks: deduct 5% of proposed holdings.

The drawback with silver is that a standard bar of 1,000 ounces is too heavy to carry and difficult to hide. And platinum and palladium are not sufficiently negotiable to fit the bill.

Never forget that precious metals are very volatile and therefore your overall investment in this entire sector should always be kept within reasonable limits. Always invest in a way which makes it possible to "average". In other words, don't spend all your money at once. Keep some in reserve so that you can purchase more if, against your expectations, the metal drops to much lower levels. Even after you have made such averaging purchases, you should still have enough money to pursue other investment opportunities and meet your obligations. As a general rule, I would say that your overall precious metals investment should never amount to more than 35% of your net worth. At the same time, I strongly recommend that you have at least 10% of your assets in fully paid gold bullion, and that you make an additional investment of between 5% and 10% either in other precious or strategic metals.

The Strategic Metals 2

Introduction

There is considerable confusion as to what a strategic metal is. Some investors believe that strategic metals are those necessary to fight wars, while others believe that any metals controlled by Soviet Russia or China should be referred to as strategic. All of these interpretations are partially true, but the best definition is as follows: *A strategic metal is one whose role in our industrial and military technology is critical, for which there are few or no substitutes, and of which the supply source is a politically unstable or hostile nation.*

There are many arguments to prove that strategic metals will fare well during the 1980's. To start with, the Seventies were a time when major western governments promoted the concept of Detente. Economic growth was relatively brisk and strategic goals were simply not pursued very aggressively. The result was that government stockpiles in Britain, Germany, France, Japan and the United States fell very low. Soon after his election, President Reagan established new stockpile targets, announcing his intention that many of the metals discussed in this chapter would be accumulated by the U.S. government. Similar action is expected from the major European nations and Japan. Obviously, such purchases will increase the demand and decrease the supply for these materials.

In addition, the Soviet/U.S. relationship has deteriorated when compared to what it was during the last decade. Looming on the horizon is the fact that the alliance between the United States and Europe is increasingly threatened by differing political viewpoints. And economic tensions, created by dwindling resource income and a rapidly escalating debt burden are causing a lot of realignments among third world nations. All this makes the world less stable and enhances the possibility of armed conflict. During the preparation for any military confrontation, the consumption of strategic metals would increase at a dramatic rate and, if war broke out, this would again have a highly positive impact on prices — assuming we are not talking about nuclear war between the superpowers.

Nevertheless, if we turn the clock back to the late Seventies which, from a global point of view, was by no means a stable period either, we find that very few people made any money in the strategic metals markets. Fortunately, we will be able to draw our conclusions and learn from the experiences of thousands of unsuccessful investors. But before I illustrate the opportunities and pitfalls of trading strategic metals, I would like to brief you on their history and function in today's world.

The History and Function of Strategic Metals

Strategic metals were as important in ancient times as they are today. Drawings from 2000 B.C. show the soldiers of Sumer trundling about in clumsy four-wheeled carts drawn by asses, while neighbouring tribes were already experimenting with primitive chariots and horses.

It was the Hittites, however, who would eventually emerge as the leading power and it was a strategic material which aided their rise. They had learned to use iron and their strategy was to use this technology to build an invincible army. Based on new weapons, such as iron-fortified chariots and iron-hooped wheels, the Hittite army became the strongest fighting force in the then-known world. They raided Syria, sacked Babylon and annexed Mesopotamia. Even the mighty Egyptians were challenged. After a fierce battle at Kadesh, Pharaoh Ramses III decided to appease the Hittites and married their king's daughter. No one could defeat them, simply because they had a metal which gave them superior technology!

When the Hittite empire finally came to an end, the use of iron spread rapidly and stimulated change not only militarily but also economically. In agriculture, iron-using tribes could now till heavy soils which had remained impervious to wood or flint. And new tools were crafted which made it possible to use softer materials, such as pulp, reeds or wood more creatively.

Today, we look back over centuries of scientific advances which make it possible to travel by air, to carry sound and light across oceans and deserts, and to cook meals in less than a minute. Most of us have at least visited a foreign continent and all of us use goods and merchandise made on the other side of the globe on a daily basis and as a matter of course.

This economic and cultural interdependence has opened up possibilities our grandparents never dreamed of. It also gives us a very false sense of security. The technology for producing our everyday goods, as well as complicated weapons systems, is now totally in the hands of industries whose survival is dependent on a vast array of metals, minerals and other natural resources. Some can be found nearby, which allows us to largely ignore this problem of dependence, but others are imported from abroad and leave us in a very vulnerable position.

Remember when we experienced a shortage of oil a few years ago?

Fuel prices rose sharply and plunged our economies into a roller-coaster ride of high inflation followed by recession. But, as painful as this adjustment was, it was only one aspect of a far greater crisis. We were forced to realize that a more serious disruption of Middle Eastern oil supplies could paralyze our entire industrial complex and bring our powerful military machine to a grinding halt.

Luckily, this experience caused a flurry of new exploration activities designed to tap North American energy reserves and, as a result, our vulnerability will ultimately be reduced. But oil is only one resource we depend on, and virtually nothing has been done to allow us to survive shortages of other materials which are equally critical to the survival of our technology.

Among these are the strategic metals in the group of non-ferrous metals. Their history, in general, is a very young one, because many were not discovered or identified as elements until a few decades ago. Each of these metals has its own definite physical characteristics, but not all of them play an important role in our modern industries. Sometimes a less expensive material with similar properties can be used, sometimes the metal is so toxic that it cannot be handled without excessive risk, and in other cases no one has yet found any useful application. On the other hand, there are some materials which are immensely important to our modern technology and indispensable to our defense.

Most strategic metals are not used individually. Instead, they are combined with two or more other metals and then employed as alloys. The most important of these are consumed in huge quantities by the steel industry in a form known as ferro-alloys. Combined with iron, individual strategic materials such as chrome, manganese, titanium or molybdenum command enough volume to have markets of their own. There are, however, thousands of other alloys which are used less frequently, but are of no less importance.

Few people realize that strategic metals are so critical that modern jet engines, T.V. and radio sets, electronic consoles, cars, submarines, explosives, even batteries could not be manufactured without them. And our dependence on strategic metals is on the increase. As technology marches on, scientists find ever more specialized ways of using each metal in an effort to improve our standard of living. Moreover, many parts of the world which previously were underdeveloped are now building up industries of their own and are rapidly joining the ranks of strategic metal users.

The Prices of Strategic Metals

There are several factors which suggest that strategic metals prices will rise for some time to come. The most important of these is that reserves of non-ferrous metals are finite. At the same time, demand is growing and the cost of production is increasing. Provided these trends do not change, the long term price outlook remains strongly in favour of strategic metals.

Political and military trends should also act to support prices. Over 60% of all non-ferrous metals used in the United States are now imported. And for more than twenty strategic metals America is totally or, to a very high degree, dependent on unreliable supply sources, i.e., Bolivia, China, India, Russia, South Africa, Thailand, Turkey, Zaire, Zambia and Zimbabwe. For some of the most vital elements, such as chromium, manganese, palladium, platinum and vanadium, the West is almost entirely dependent on just three countries: South Africa, the Soviet Union and Zimbabwe.

There is little doubt that the long term goal of the Russians is to isolate mineral-rich South Africa by means of an encircling movement which, to a large extent, is already accomplished. Angola and Mozambique, which only a few years ago were Portuguese colonies, are now Russian satellites stocked with Cuban soldiers. Rhodesia, another of South Africa's former allies, recently elected a black government and, under its new name Zimbabwe, has entered into a strong relationship with the Soviets. If the Russians ever succeeded in controlling South Africa, the consequences to the West would be disastrous. In addition to a monopoly over the metals and minerals of Southern Africa, they would also be in command of the oil tanker routes between the Persian Gulf and Europe and North America. Our industrial and military survival would then truly be in the hands of the U.S.S.R.

It is inconceivable that the West would allow this to happen without a fight. But that does not mean the Russians will stop trying. The probability is therefore high that the southern tip of Africa will some day become an area of political tensions, which means that the supplies of strategic metals would be threatened. It is easy to imagine what this would do to prices.

In principle, the Western allies have agreed that they should prepare for this eventuality by stockpiling, but they have simply not had the resources to carry through. It is a shocking paradox that the greatest threat to Western democracy may be our own democratic principles. If, during the recent economic slump, strategic metals purchases had been given priority over unemployment or welfare, the government would have found itself out of power very quickly.

During the late Seventies, when the economy was better, several Western nations reviewed their stockpile policy and vowed to improve the situation. West Germany, for instance, decided to spend over $350 million to help its industries acquire adequate supplies of asbestos, chromium, cobalt, manganese and vanadium. The government did so because experts had calculated that if only 30% of the annual chromium imports into West Germany were suspended, the entire gross national product of the country would fall by a full 25%! In the United States, a major study of the nation's dependence on critical materials was completed under President Carter. Realizing that the United States was now far more dependent on strategic metals than on oil, government officials were alarmed enough to introduce stockpile revisions similar to those made in West Germany. Under President Reagan, this policy was formulated further. Regrettably, the West Germans, the Americans and other Western nations have so far done very little to improve their stockpiles, primarily because dealing with the recession took precedence. Eventually, however, this issue will have to be dealt with. *The longer the present situation drags on, the more pronounced the final crisis will be.*

But, if prices were to explode due to these factors or as a result of political tensions, this would have remarkably little effect on consumption patterns. This is because the demand for strategic metals is rather "inflexible". In most applications, an individual metal or alloy forms only a minute but very important part of a particular process or device. Thus, the value of the metal used represents an insignificant part of the overall unit cost. In other words, as we saw in the case of silver, the price of the metal can still increase several hundred percent without having a noticeable impact on sales. For instance, modern car engines contain catalytic exhaust converters which are made with platinum and palladium alloys. If the price of palladium and platinum tenfolded, it would become much more expensive to replace the converter in your car, but the price tag of the entire automobile would change by no more than $200. Overall car sales would suffer very little.

Nevertheless, when combined, these political and economic factors could have an extraordinary impact on the prices of individual strategic metals. But before we go on to exploring which materials are the most interesting to the investor, I should like to make a few comments on some of the other factors which affect prices, and on the unusual structure of the strategic metals market which, quite often, can influence the price all on its own.

Most industrially used materials are traded in open markets and usually it is an exchange which provides the trading mechanism. Strategic metals are not traded in this way. Because very few transactions take place, most of the time they are arranged directly between the producer and the industrial user. In some cases, brokers or merchants act as middlemen, but this has usually no bearing on the price. As a result, the price of a specific strategic metal is determined by how badly a particular user needs it and how ready a producer is to let him have adequate supplies. So far, we have discussed the political and economic trends which will influence demand and supply in the Eighties, but more traditional price factors, such as strikes, environmental considerations, freight and storage rates are also important. And because the market is so dependent on just a few users and producers, one aspect peculiar to strategic metals prices cannot be ignored: the judgement of one particular producer, merchant or industrial user can have an overriding impact on the price.

Take, for instance, the market for selenium, a metal widely used in the manufacture of transformers, semi-conductors and photoelectric cells. Because the transfer of a photographic image by means of static electricity can best be engineered by using this material, selenium is in great demand in the photocopying industry. But the metal's total annual production is only 1,000 metric tons, and its current price is a mere $12 per pound. The entire world output, therefore, could be bought for less than $30 million, which means that an individual investor or industrial user could easily corner the selenium market in order to control or bring to a stop the photocopying industry.

In reality, it would be very hard to actually buy up all of the world's selenium, precisely because the market is so small that everyone would soon know about such a plan. But the purchase of even a portion of the world's selenium reserves would suffice to drive up the price considerably. This situation is not unusual for strategic metals. Many of the materials we will discuss represent relatively small markets and, as a result, fluctuate sharply.

Another factor contributing to price volatility is the inflexibility of demand which we discussed earlier. Because the typical application of a strategic material is its use as a very small part of a much more sizeable unit, the price can rise sharply without affecting sales patterns. When demand is high, producers are unable to increase mining output, but consumers are not overly concerned with the higher price they have to pay. During the volatile Seventies, for instance, cadmium, cobalt, molybdenum and indium advanced by more than 500% during times of high demand. However, the recent global recession reversed these trends and most strategic metals saw equally sharp declines. Not only were speculators and holders of large inventories punished by high interest rates, but industrial demand also dropped.

In the long run, however, it is likely that countries supplying critical materials will become more aggressive. China, for example, has started to use its advantage as a major producer and occasionally influences the world price of antimony and tungsten. Zambia and Zaire have done their best to keep the price of cobalt high since the 1978 invasion by rebel forces into Zaire created a shortage. If it were not for the fact that some of these countries have debt repayments to make, they would probably not be willing to sell their strategic metals so cheaply. The large mining conglomerates also have an interest in keeping prices high and may, through their international trading subsidiaries, begin stockpiling these materials in order to absorb excess supply. Finally, some people even predict that sizeable producer cartels will be formed in the not too distant future. With this, however, I must disagree. The objectives of the producing nations are simply too different — as, they are, for instance, in the case of South Africa and Russia.

But, to the investor, all this matters very little. What is of far more importance is that right now strategic metals prices are depressed and must go up. What is a shocking and alarming crisis to the Western industrial and military complex, could be a unique opportunity for you.

Two Precious Metals With Outstanding Strategic Qualities

In our chapter on platinum and palladium, you learned that these metals greatly increase the supply, and reduce the cost, of the world's foods and fuels. In addition, applications in the defense, transportation, communications and medical industries cause platinum and palladium alloy catalysts to be used in nearly twenty percent of all manufactured goods.

Unfortunately, for the West, their sources of supply are not very reliable. The major producer of platinum is South Africa with 65 percent, followed by the Soviet Union. In the case of palladium, Russia is the major producer with 70 percent, while South Africa is second. Although Canada's share in each case is the third largest, its contribution of around five percent could not possibly supply world demand if Russian and South African shipments were interrupted.

As you already know, the demand for platinum and palladium is centred largely in Europe, Japan and the United States. This imbalance has caused many experts to predict a catastrophe if Russia ever denied the West the metals it needs to support its industries. And what about South Africa? A recent policy background paper published by the government of Ontario (where most of Canada's platinum is found) summarizes the problem: "A denial of oil to the economy of South Africa, civil disorder in South Africa, or an invasion, could each seriously disrupt platinum output and push up the world price." And it continues: "It would take an estimated ten years for the U.S., and probably other countries, to adjust to the cutoff."

Canada is not alone in recognizing this danger. Shortly after his election, President Reagan made the following statement introducing measures to restructure the national defense stockpile of strategic and critical materials: "It is now widely recognized that our nation is vulnerable to sudden shortages in basic raw materials that are necessary to our defense." The White House announced that it would initially purchase $100 million worth of such materials, and that purchase priority would be given to cobalt. Along with it, nine other critical materials were listed as candidates for replenishment. It should come as no surprise to learn that platinum-group metals were included.

These government statements illustrate the West's vulnerability and its dependence on platinum and palladium from the Soviet

Union and South Africa. Yet, current U.S. stockpile levels are still far short of the targets established in the late Seventies, even before President Reagan came on the scene. There are now 0.45 million ounces in Washington's platinum hoard, although 1978 targets call for 1.31 million ounces. In the case of palladium, some 2.54 million ounces are needed, but only 1.25 million ounces are there! It will shock you even more to learn that virtually none of the stockpile inventories of platinum and palladium meet current requirements for purity nor are they in the form which suits U.S. defense industry needs.

The low level of western stockpiles, their critical importance to our defense, petroleum and agricultural industries, the geographic and political imbalance between producers and consumers, and the fact that for most industrial applications they cannot be readily substituted, give platinum and palladium the advantage of being the only precious metals which are strategic as well. For the reasons discussed in this chapter, not to mention the other factors which were reviewed in the first part of this book, I have to conclude that the case for investing in these two strategic, precious metals is overwhelming.

Twenty-Four Other
Strategic Metals Analysed

Before we concern ourselves with how to invest in strategic metals, we have to apply the same criteria we used when looking at platinum and palladium to the many other strategic metals which at first glance seem suitable.

If you look at the chart on page 148, you will notice that the third metal listed is cadmium, and that it is used in the marine, aerospace, chemical, automotive, optical and explosives industries. I am not very intrigued by the potential for cadmium because in most of its uses it can be readily substituted. Also, you will notice that the U.S.S.R. is responsible for seventeen percent of production, but Japan and the United States each supply nearly as much. Therefore, the producer/user situation is balanced. There is only one reason why the cadmium price might advance, and that is that the U.S. stockpile is very low. But, for my own purposes, one positive reason is not enough — not when there are several strategic metals which fulfil all our criteria for appreciation.

A much more satisfactory example would be cobalt, the fifth metal from the top in our table on page 150. Cobalt alloys are essential to our electronics and aircraft industries. A modern jet engine cannot be manufactured without them and many electro-magnetic components depend on the metal. Moreover, there is almost no substitution possibility. Well over a third of the world's cobalt is used in the United States, but most production is in Zaire (58 percent), the Soviet Union (9 percent), and Zambia (8 percent). Even worse, or better from an investment point of view, the cobalt stockpile in the United States is still quite low. This strategic metal, therefore, should do particularly well because it satisfies all of our objectives.

If we use this method of selection for all strategic metals, we arrive at three which look particularly promising and, as you can see, I have marked them on the tables with an arrow. They are germanium, cobalt and tungsten. There is also an entire category of metals which qualify for only two of our three criteria, and many of these also have intriguing price possibilities. The ones I am most attracted to are chromium, manganese, tantalum and vanadium.

You should realize, however, that there are some other factors not taken into account in my tables. For example, I did not consider whether new applications might be found in the future, because such an analysis would require far greater understanding of the latest

STRATEGIC METALS RESEARCHED

APPLICATIONS	GROUP AND METAL	% OF GLOBAL PRODUCTION	SUBSTITUTION
automotive, chemical, greases, ceramics, metal refining	CHEMICAL METALS		
	ANTIMONY	China 30 USSR 17 USA 7	considerable
	LITHIUM	Brazil 14 USA 32	little
marine, aerospace, chemical, automotive, optical, explosives	ELECTRONIC METALS		
	CADMIUM	USSR 17 Japan 16 USA 13	considerable
	INDIUM	Japan 27 CDA 20 USSR 17 USA 7	little
	MERCURY	USSR 25 Spain 17 Italy 12 China 10	some
	SELENIUM	Japan 38 CDA 28 USA 15	little
	TELLURIUM	USA 36 Japan 29 CDA 22 Peru 12	some
aerospace, marine, nuclear, electrical, chemical	NUCLEAR METALS		
	BERYLLIUM	USA Brazil Comecon **	little
	ZIRCONIUM	Australia 92	little
electronics, optics, measuring equipment, chemical	OTHER METALS		
	GALLIUM	SWITZ 63 USA 30	little
	↘ GERMANIUM	Zaire 30 USA 19 Comecon 21	little
	SILICON	USA 27 USSR 24 Norway 13	little

** Neither the U.S. nor the Comecon Nations divulge production figures for Beryllium.

PRODUCER/USER IMBALANCE*	U.S. STOCKPILE STATUS	PHYSICAL TRADING MARKETS & LIQUIDITY
moderately negative	high	*reasonable*: active producer and merch. activity
positive	low	*poor-reasonable*: producer dominated
balanced	very low	*reasonable*: open market with merchant activity
balanced to positive	low	*reasonable*: active merchant participation
negative	high	*poor*: users determine price on bid-system
positive	low	*poor-reasonable*: little merchant activity
positive	low	*poor*: producer dominated market
balanced	low	*very poor*: tight producer market
moderately negative	low	*very poor*: almost no secondary market
positive	low	*poor*: producer controlled market
negative	low	*poor*: producer dominated market
balanced to positive	adequate to high	*reasonable*: producers and merchants active

*Producer/User Imbalance takes into account current political relations.

STRATEGIC METALS RESEARCHED

APPLICATIONS	GROUP AND METAL	% OF GLOBAL PRODUCTION	SUBSTITUTION
	LIGHT METALS		
aerospace, electronics,	ALUMINUM	USA 30 USSR 16 Japan 7	some
automotive, nuclear, chemical	MAGNESIUM	USA 38 USSR 30 Norway 16	some
	TITANIUM	USSR 49 USA 30 Japan 14 UK 5	little
	STEEL INDUSTRY METALS		
aerospace, turbines,	CHROMIUM	USSR 24 SAFR 23 Albania 9	little
electronics, petrochemical	◢ COBALT	Zaire 58 USSR 9 Zambia 8	little
pipelines, automotive, nuclear	COLUMBIUM (or NIOBIUM)	Brazil 75 CDA 8 Nigeria 5	Vanadium in most cases
mining equipment,	MANGANESE	USSR 35 SAFR 23 Gabon 9 Austr. 8	little
batteries, ammunition	MOLYBDENUM	USA 57 CDA 17 Chile 11 USSR 10	little
	NICKEL	USSR 22 CDA 21 Japan 12	little
	TANTALUM	NIgeria 51 CDA 10 Brazil 8 Rwanda 8	much
	◢ TUNGSTEN	China 23 USSR 20 Bolivia 7	little
	VANADIUM	SAFR 46 USA 18 USSR 15	some

PRODUCER/USER IMBALANCE*	U.S. STOCKPILE STATUS	PHYSICAL TRADING MARKETS & LIQUIDITY
positive	adequate	*excellent*: cash and futures at LME
balanced	low-adequate	*poor*: producer dominated little merchant activity
balanced	very low	*poor-reasonable*: regular merchant activity
highly negative	adequate	*reasonable*: active producer and merch. activity
highly negative	very low	*reasonable*: prod. and merchants very active
moderately negative	low-adequate	*poor*: producer market, little merchant activity
negative	adequate	*reasonable*: active producer and merch. activity
positive	very low	*poor-reasonable*: merch. very inactive.
balanced	very low	*excellent*: cash and futures at LME
negative	low	*poor:* producer dominated little merchant activity
highly negative	low-adequate	*reasonable*: merchants run an active market
moderately negative	very low	*poor-reasonable*: producer dominated.

scientific advances than I have access to. Also, I was not in a position to accurately assess how much of each metal is stored by producer nations, mining companies and merchants. The Western governments publish official stockpile figures, but private dealers and state run mining conglomerates, like those in Russia or China, do not.

Finally, you should realize that buying strategic metals can be a long term affair. Just because a negative producer/user balance, low possibilities of substitution and low government stockpile figures all unite to enhance the price outlook of a metal, it does not follow that your investment will go up in the near future. Poor economic conditions, reduced inflation and political stability could delay the inevitable price explosion for years to come.

Investing in Strategic Metals

Physical Material

In today's volatile world, negotiability is by far the most important aspect of any investment. What good can it possibly do you if the price of something you own has just zoomed up by a few hundred percent and you are unable to sell it? Unfortunately, liquidity can sometimes be a problem with strategic metals investments. Platinum and palladium are the only two strategic materials which are also precious and which are, therefore, invariably negotiable.

Some of the other strategics are used primarily by industrial interests, and are not suitable for the investor who wants to take personal delivery. In the case of chromium, for instance, you would get your metal in the form of chromite sold by the ton, or in the form of ferro-chrome packed in heavy steel drums. If you invested in cobalt, you wouldn't even want to take delivery, because the metal comes in the form of highly toxic cobalt cathodes, and cobalt granules. Luckily, there are warehouses which can store these materials for you. Simply take delivery there.

A number of other strategics, including germanium (one of the metals I recommend) are much easier to handle. They are traded in ingots of high purity which could even be stored in your home — provided your home were big enough. Bear in mind that these materials are usually sold in very sizeable quantities. An investor wanting to buy vanadium, for instance, would have to purchase ten tons, worth about $60,000!

TYPICAL LOT SIZES FOR STRATEGIC METALS

Antimony	10 tons	Manganese	100 tons
Beryllium	1 ton	Mercury	50 flasks
Cadmium	5 tons	Molybdenum	10 tons
Chromium	5 tons	Selenium	1 ton
Cobalt	1 ton	Silicon	5 tons
Columbium (Niobium)	5 tons	Tantalum	1 ton
Gallium	100 kilograms	Tellurium	1 ton
Indium	100 kilograms	Titanium	5 tons
Iridium	100 ounces	Tungsten	25 tons
Lithium	10 tons	Vanadium	10 tons
Magnesium	20 tons	Zirconium	5 tons

The most efficient and least expensive way to buy strategics is to go directly through a London broker. There, strategic metals are often referred to as "minor", to differentiate them from the major industrial metals, such as copper, zinc, nickel, lead or tin. The firms specializing in these materials are members of The Minor Metals Trading Association, an exclusive group of about seventy brokers, dealers and producers. Among them are companies which are very specialized in the metals sector, such as Ametalco Trading, Brandeis Goldschmidt, Intsel, Leopold Lazarus or Lonconex, and which also hold seats on the London Metal Exchange and on New York's Commodity Exchange.

The reason why London should be used for such transactions is twofold. To start with, London provides the only market mechanism for many of the strategic materials. And, secondly, London brokers are the only ones who have traded in these metals for many years and whose knowledge in this field is undisputed.

But remember, buying strategic metals is only half the battle. You still have to sell them and, unfortunately, there are times when neither merchants nor industrial users will want to quote you a price. At first, this may seem surprising but, if you think about it, it is very logical. An industrial user of chromium in the United States will obviously not go to a South African exporter on a weekly or monthly basis to secure his supplies. Instead, he will purchase in great bulk to take advantage of the economy of scale and thus reduce his transportation charges. Ideally, he will purchase when interest rates are low and liquidity conditions are such that he can easily get the necessary bank credit to finance his stockpile.

The inevitable result is that at certain times there is dramatic demand for such materials while, at other times, particularly when prices seem excessive, demand totally dries up. These factors are beyond any investor's control — as some who tried it in the Seventies found out to their sorrow.

But don't lose sight of the fact that your strategic investment has little to do with wanting to exploit the inflation cycle. If you want to do that, trade gold instead. The reason for buying a strategic metal is that sooner or later a dramatic shortage will be caused by a supply cut-off or massive government stockpiling. Your patience may never be rewarded, but if it is, you will be one of the very few owners of a resource everyone wants!

Partial Lots

During the late 1970's, when strategic metals appreciated almost daily, several British and American investment dealers tried to make the market more accessible to individuals. They bought standard lots of materials and had them stored in warehouses in Amsterdam and London. With a mark-up, they then sold smaller units of their hoard to investors, particularly in North America.

This approach seemed entirely practical but, as many unfortunates soon found out, their dealer all too often could not find a buyer when they wanted to resell their investment. When push came to shove, the market for partial lots proved to be as illiquid as the market for the underlying material, precisely because small quantities were involved.

Mining Shares

Shares of producing mining corporations provide some investment possibilities for strategic metals, although there are a great many complications. Most companies produce one or more strategic metals along with all kinds of other ores and minerals, which makes a direct participation very difficult. A typical case is Amax Incorporated, an American mining company listed on the New York Stock Exchange. Amax has everything you want in a share investment, from an established dividend record to sound management. But one share in that company is equivalent to 12 tons of molybdenum, 3.7 tons of copper, 7.4 tons of iron ore, 54 tons of coal, 9.1 tons of phosphates, 0.4 tons of tungsten, .37 tons of lead and zinc, 1.2 tons of potash and .13 tons of various other ores. Obviously, Amax shares are not a great tungsten play, but represent excellent diversification in the metals sector. Another example in this context is Falconbridge Nickel, also one of the world's most important mining and resource companies. Its products include nickel, copper, cobalt, gold, silver, platinum group metals, selenium, lead, iron ore, zinc, cadmium and more. The company's dividend record is good and their shares are listed on the Toronto Stock Exchange and on NASDAQ. But Falconbridge shares are obviously of little use if it is cobalt you wish to invest in. Even if the cobalt price tenfolded, your investment would not benefit to a major degree, because cobalt is only produced in small quantities as a by-product of other ores.

But what can you do if you want to invest in one of the metals we singled out as the most likely for appreciation? Again, platinum and palladium provide us with a relatively easy answer. Both Impala Platinum Holdings and Rustenburg Platinum Mines, the companies we discussed in the chapter on precious metals mining shares, represent a direct investment in platinum and palladium and are traded in New York, London and Johannesburg.

Tungsten, one of the three metals we singled out previously, is another intriguing possibility. The largest free world tungsten interest is in Canada, is managed by Canada Tungsten Mining Corporation Limited, and listed on the Toronto Stock Exchange. Canada Tungsten is a conservative and established company which offers one of the few direct plays in an important strategic material.

Of the second group of metals we targeted in our analysis, manganese and chromium are the best candidates. A South African firm, S.A. Manganese Amcor Ltd. ("SAMANCOR"), offers a relatively direct investment in both of them simultaneously. Listed on the Johannesburg Stock Exchange, SAMANCOR is a highly regarded mining firm with a long dividend record. While it is a major producer of chromium ferro-alloys, it is also the world's largest exporter of manganese.

Unfortunately, there aren't many other mining share companies for strategic materials. The reason for this is the fact that those mines which produce strategic metals are in countries which either do not have stock markets or whose shares cannot be traded abroad because of currency controls. In other words, most strategic metals are mined in countries which are inherently unstable — which is why we were worried about supplies in the first place!

Funds and Managed Accounts

Reflecting on the experience of hundreds of investors who were unable to sell their small strategic metals holdings as prices collapsed, a number of experts predicted that "unit trusts", or mutual funds, would become the vehicle of the future. And, from what we have seen in this chapter, I have to agree that this is the best way to go.

A strategic metals fund, after all, can eliminate many of the disadvantages of individual investment. To start with, it allows a group of investors to "act" as one, thus eliminating the problem of dealing in small size. Because different participants in a fund have different objectives, the managers don't always have to go to an outside source to fill an order — one of the fund's own clients may be buying while someone else is selling.

Funds can also give you diversification. Without a cash commitment of at least $200,000, you simply could not buy a diversified strategic metals portfolio in tradeable sizes. Buying certificates in only one material, on the other hand, leaves you wide open to the risk that a less expensive substitute may suddenly be found, or that a newly discovered alloy can do the same job better. A mutual fund can give you a direct interest in a variety of metals for a relatively small investment.

The best known funds are those run by London metals brokers and some are quoted daily in the London Financial Times. The minimum required to participate ranges from $10,000 to $25,000 and fees vary depending on the size of your investment and, sometimes, on the performance of the fund. In most cases, they are relatively high: entrance fees of around ten percent are not uncommon, and the transaction and management fees charged to the fund during any one year frequently exceed that figure. But you have to see this fee structure in the right context. Take into account that the business of running a strategic metals portfolio is a highly specialized one and consider these fees as the price at which you acquire a manager's expertise. Also, you should not forget that when the time is right the upside potential for these metals is simply enormous, provided you are holding the right one!

Two London firms which offer various strategic metals funds are Hargreaves & Williamson and Strategic Metals Corporation. Another firm, Strategic Metal Trust, is registered in the Isle of Man and administered from Switzerland. The inventories of all three funds are held in depositories approved by the London Metals Exchange in London or Rotterdam and are fully insured. I suggest you obtain details by writing to these firms and comparing their programs. Their addresses are contained in our Guide to Banks, Brokers and Dealers on page 167.

Another way to avail yourself of someone else's expertise is to enter a managed account arrangement. Hargreaves & Williamson, for instance, offers two types of managed account plans. The first allows you to choose a single metal which is then bought and sold at what the manager regards as opportune times, while the second allows you to participate in a balanced metals portfolio. The portfolio managed by Strategic Metals Corporation operates similarly: with a minimum of $10,000 you have access to a basket of up to forty industrial metals of the manager's choice.

The people who run Strategic Metals Trust, Troy Associates of Geneva, also offer managed account facilities, but theirs is really an investment in metal shares with an emphasis on strategic materials. The same concept is used by AAA Consultants Limited of the Cayman Islands, where a minimum investment of $50,000 is required. Although defined more generally as a "resource share portfolio", strategic metals represent a reasonably high component of overall holdings. The AAA consulting group, by the way, includes a number of experienced and well-known analysts, such as Pamela and Mary Anne Aden, Alexander Paris, and Robert Meier.

A significant number of new funds and managed account plans are expected to begin operations within the next two to three years, a development which will give investors an even greater freedom of choice.

Financing Strategic Metals

The rationale behind financing strategic metals is exactly the same as discussed in the chapter on precious metals. Buying on margin or arranging a collateral loan, in other words, will free up some of your money which can then be utilized for other investments or to increase the size of your already existing investment in strategic metals.

There are, however, some distinct differences in the way lenders behave towards strategic metals. First of all, strategic metals are not well known. Even if a banker has heard of titanium, it is very unlikely that he also knows what it is and what affects its value. This means that he will not want to lend you money to invest in strategic metals nor will he accept it as collateral against a loan. How can you by-pass this problem and exploit the opportunities in the strategic metals sector more efficiently?

The most effective and inexpensive way of doing so, quite obviously, is to use something else as collateral with your bank. In most cases, the tax advantages of having borrowed for investment purposes are just as advantageous as they are with margin loans. Besides, if you put up some of your shares, bonds or real estate holdings as collateral, the interest rate will probably be better anyway. Once the money is advanced, you can go to a strategic metals broker and buy the material outright.

The only drawback to this solution is that most investment loans are in the demand category. In other words, the bank can ask you for its money back at any time, although this is not often done. Nevertheless, it is a risk you should take seriously enough to provide for.

There are also some metals brokers who provide financing, particularly on larger transactions. But don't be surprised to find that the leverage you can get is smaller than it would be, let's say, for gold or for some other widely traded commodity. Because strategics have a history of very volatile price behaviour, the lending risk is higher and the suitability of strategic materials as collateral is therefore lower.

A specialized dealer might also point out that one of the best financing possibilities is provided by the market mechanism itself. Most materials in the category of strategic metals can be traded for "forward delivery", provided the transaction is sizeable enough. This means that a deal can often be arranged for a firm price with a producer while the actual delivery is deferred into the future.

Obviously, your dealer would have to pay a premium over the going cash price for the metal and this would reflect, more or less, the cost of money until delivery date. In other words, instead of paying interest to a lender, you purchase the metal in question at a higher price. As brokers do with futures contracts, the metals dealer would ask you for an initial downpayment and, whenever market declines occurred, would require additional deposits.

Finally, another way of buying strategic metals "on margin" is of special interest to those who want to buy units in a fund or who want to open a managed account. Before selecting a plan, ask the manager whether he can help you leverage your investment. Quite often, such firms have a prior arrangement with a specific bank which, in return for all their business, specializes itself in loans of this kind. Although relatively few funds and accounts managers encourage you to borrow against the investment you hold in their plan, quite a number of them will gladly make it possible for you to invest more money in their vehicle than you actually put up yourself.

Storing Strategic Metals

As we have seen, the majority of strategics are traded in sizeable units and seldom in a form suitable for storage by yourself. In practice, most metals held for investment are never actually delivered — but are instead traded between dealers who use receipts or documents evidencing their storage at an approved warehouse.

All exchanges have a list of approved warehouses and exercise great care in their selection of suitable depositories. In order to get onto that list, the owner of a warehouse has to comply with certain security and insurance standards which are then constantly audited by the exchanges and, in most cases, the regulatory authorities.

That is why it makes sense to carefully check the purchase confirmation received from your dealer, in order to ascertain where the metals are kept. Do not allow your dealer to keep them at a bank, no matter how reputable. Be sure that the warehouse is named and that it is one of those which are approved by a major exchange. This guarantees that your metals are safe and also makes selling them much easier. For example, if your metals are held at an exchange approved warehouse and your dealer cannot quote you a competitive price, all you have to do is get your storage receipt transferred to your own name. Your metals can then be sold to any of the other dealers who use the same facility. It is far less likely, however, that a transaction could be arranged if your materials were held at a bank where only your own dealer did business.

Exceptions to this rule, of course, are platinum and palladium. In industrial forms such as plate or sponge, both metals are more negotiable when held at exchange approved warehouses. But bars and wafers of a high purity are traded almost exclusively between the bullion dealing banks, which means that a vault is the right place to store them.

Following The Markets 3

Newsletters and Advisory Services

Most investors subscribe to one or more financial newsletters. Following these can be extremely useful but it can also have disastrous consequences. As a professional involved in actively trading precious and strategic metals, I consistently read well over a hundred newsletters from around the world. This routine has taught me many lessons, but one stands out above all others: unless you can use a newsletter to your advantage, it will do you no good at all.

Let me explain this further. Every article appearing in a newsletter or through an advisory service is authored by one particular analyst who has his own strengths and weaknesses. If you follow these people for a long period of time, you get to know them almost as well as your neighbours or your colleagues at work. Some have a tendency to exaggerate, others have a habit of covering up their mistakes or are consistently self-congratulatory. You also get to know which pet theory each analyst is in love with. Very few people have the ability to quickly adjust their thinking to new developments. Some are convinced there will be a deflation, while others have predicted hyper-inflation for the past 35 years. Some analysts are consistently cautious and responsible, others always live in worlds of the extreme.

The safest way of using newsletters is to be very skeptical and never to act on the advice of one particular analyst if you haven't been following his recommendations for at least six to twelve months. Subscribing to more than one newsletter gives you added security. You will be able to compare one expert's ideas to the results of someone else's analysis. Keep a record of your favourite analyst's recommendations and, alongside, write down whether you agreed with their conclusions at that time. Check later how accurate their advice was and where your own analysis of the situation went wrong, or why it was right. After a while, you will develop a fairly good understanding of how useful the information contained in various newsletters is. That is when you can safely discontinue the bad ones and instead try out some services you haven't yet explored.

During this process, you should try to get to know yourself as well. Don't be influenced too easily — many newsletters are written in an excellent and convincing style, which does not mean that their recommendations are sound. If you are the type who gets excited during political speeches, be twice as careful!

Finding the right combination of newsletters or advisory services will initially be a frustrating, time-consuming and expensive task. However, if you stick it out you will eventually find one or two advisors who are flexible, responsible, and whose recommendations and guidance you trust. That is when the newsletter investment will pay you back many times over. If they are used for investment purposes, by the way, you may deduct subscription fees from your taxable income.

Out of the many newsletters and advisory services I know, I find the ones listed below to be by far the most useful:

Mining Journal, published by The Mining Journal Limited, 15 Wilson Street, Moorgate, London EC2M 2TR, England.
A professional's guide to the latest developments affecting metals and minerals. The only periodical containing financial results and mining statistics affecting all South African gold mines. Also features the latest prices for the many strategic metals, and other materials. Weekly.

Green's Commodity Market Comments, P.O. Box 174, Princeton, New Jersey 08540.
Specializing in gold and silver, this letter is among the most respected by professional traders. Green's gives in-depth coverage of the latest supply and demand developments and recommends trading and investment strategies. Bi-weekly.

The International Bank Credit Analyst, published by BCR (Boeckh, Coghlan Research) International Publishing Ltd., 3463 Peel Street, Montreal, Quebec H3A 1W7.
An in-depth study of economic trends around the world, with technical and fundamental analyses of precious metals, currencies, interest rates and stock market trends. Monthly.

Friedbergs Commodity and Currency Comments, Hume Publishing Limited, 4141 Yonge Street, Willowdale, Ontario M2P 2A7.
A highly respected analysis of fundamental technical trends in the areas of precious metals, currencies and interest rates. Monthly.

The Metals Investor, 711 W. 17th St., G-4, Costa Mesa, CA 92627.
An excellent in-depth analysis of strategic and precious metals trends. Features latest prices for strategics. Monthly.

Gold Newsletter, published by the National Committee for Monetary Reform, 4425 West Napoleon Avenue, Metairie, LA 70001.

A compilation of the latest views on gold. Gold Newsletter regularly prints articles by professional gold watchers, ranging from economists to brokers, from the "gurus" to conservative bankers. Also reports on new markets, new investment vehicles and related developments. Monthly.

Silver and Gold Report, published by Precious Metals Report Inc., P.O. Box 325, Newtown, CT 06470.
Interviews with leading precious metals analysts, market surveys, etc. Semi-monthly.

Deliberations, published by Ian McAvity, P.O. Box 182, Adelaide Street Station, Toronto, M5C 2J1.
The leading letter on technical chart analysis for precious metals, currencies and stock market trends. Semi-monthly.

Aden Analysis, 4425 West Napoleon Avenue, Metairie, LA 70001.
Technical analysis of precious metals and other financial markets, using an unusually broad range of comparisons and statistics. Semi-monthly.

My own recommendations are contained on a regular basis in the following two newsletters: Personal Finance and International Investment Letter, both published by Kephart Communications Inc., 1300 N. 17th Street, P.O. Box 9665, Arlington, VA 22209.

Guide to Banks, Brokers & Dealers

AAA Consultants Limited
P.O. Box 472
Grand Cayman
Cayman Islands

Funds manager; specializes in precious metals and resource shares.

A-Mark Precious Metals Inc.
9696 Wilshire Boulevard
Beverly Hills, Cal. 90212-2378
United States

Regional precious metals retailer.

Ametalco Trading Limited
29 Gresham Street
London EC2V 7DA
Great Britain

International metals dealer.

J. Aron & Company
160 Water Street
New York, N.Y. 10038
United States

Major international metals dealer; Official distributor U.S. gold program and most bullion coins.

Ayrton Metals Ltd.
Imperial House
Dominion Street
London EC2M 2SD
Great Britain

Precious metals dealer.

Bache Halsey Stuart Inc.
100 Gold Street
New York, N.Y. 10038
United States

Major international metals dealer and brokerage firm; retails bullion, certificates, futures, options, mining shares.

Bank of Delaware
300 Delaware Avenue
P.O. Box 791
Wilmington, Delaware 19899
United States

Issues registered and bearer certificates for gold and silver.

Bank Julius Bär & Co. AG
Bahnhofstrasse 36
CH 8022 Zurich
Switzerland

Precious metals dealer and retailer.

Bank Leu
Bahnhofstrasse 32
CH 8022 Zurich
Switzerland

Major international bullion and numismatic dealer.

Bank of Nova Scotia
44 King Street West
Toronto, Ontario M5H 1H1
Canada

Major international bullion dealer; issues gold, silver certifiates.

Brandeis Goldschmidt & Co. Ltd.
4 Fore Street
London EC2P 2NV
Great Britain

International metals dealer.

BGR Precious Metals Inc.
330 Bay Street
Suite 1403
Toronto, Ontario M5H 2S8

Precious metals investment corporation.

Canadian Imperial Bank of Commerce
Main Branch
Commerce Court
Toronto, Ontario M5L 1H1
Canada

Issues gold, silver certificates.

W.I. Carr Sons & Co.
2 Ice House Street
St. George's Building
8th Floor
Central, Hong Kong
Hong Kong

Hong Kong bullion dealer.

Citibank N.A.
Gold Center
399 Park Avenue
Sort 0907/Branch Lower Level
New York, N.Y. 10043
United States

Issues gold and silver certificates.

Credit Suisse
Paradeplatz
CH 8021 Zurich
Switzerland

Major international bullion and coin dealer; member of Zurich gold pool.

Deak Perera Inc.
29 Broadway
New York, N.Y. 10006
United States

Precious metals retailer; issues gold, silver, platinum and palladium certificates.

Dean Witter Reynolds Inc.
#5 World Trade Center
New York, N.Y. 10048
United States

Major brokerage firm; sells bullion, coins, mining shares, futures and options.

Derby & Company Ltd.
Moor House
London Wall
London EC2Y 5JE
Great Britain

Major international bullion dealer.

Deutsche Bank A.G.
18 Rossmarkt
D-6000 Frankfurt am Main
Germany

Precious metals dealer.

Dominion Securities Ames Ltd.
P.O. Box 21
Commerce Court South
Toronto, Ontario M5L 1A7
Canada

Major brokerage firm; sells bullion certificates, mining shares, futures and options.

Dresdner Bank AG
Gallusanlage 7-8
D-6000 Frankfurt am Main
Germany

International bullion and coin dealer.

Dreyfus Gold Deposits
600 Madison Avenue
New York, N.Y. 10022
United States

Gold deposit receipts.

Drexel Burnham Lambert Inc.
60 Broad Street
New York, N.Y. 10004
United States

Major international brokerage firm; sells bullion and coins, mining shares, futures and options.

Dynamic Funds Management Limited
330 Bay Street Suite 1403
Toronto, Ontario M5H 2S8 Canada

Dynamic - Guardian Gold Fund; other no load funds.

The First National Bank of Chicago
One First National Plaza
Chicago, Ill. 60670
United States

Gold savings passbooks; precious metals storage accounts.

Gold Plan AG
Volkmarstrasse 10
Box 213-02
CH 8033 Zurich
Switzerland

Gold accumulation plans; Gold insurance plans.

Guardian Safe Deposit Inc.
2499 North Harrison Street
Arlington, VA 22207
United States

24-hour private safekeeping facilities.

Guardian Trust Company
74 Victoria Street
Toronto, Ontario M5C 2A5
Canada

Bullion dealer; issues gold, silver, platinum and palladium certificates; represents British Royal Mint.

Hang Seng Bank Limited
77 Des Voeux Road Central
Hong Kong
Hong Kong

Bullion and coin dealer.

Hargreaves & Williamson Ltd.
201 Borough High Street
London EC1A 9HN
Great Britain

Strategic metals broker; unit trust; managed accounts.

Hong Kong Bank
1 Queen's Road Central
Hong Kong
Hong Kong

International bullion dealer; issues gold certificates.

E.F. Hutton
One Battery Park Plaza
New York, N.Y. 10004
United States

Gold and silver accumulation accounts.

Intsel Ltd.
83-87 Gracechurch Street
London EC3V 0AA
Great Britain

International metals dealer.

Johnson Matthey Bankers Ltd.
5 Lloyds Avenue
London EC3N 3DB
Great Britain

Major international bullion dealer; member London gold market.

Leopold Lazarus Ltd.
Gotch House
20-34 St. Bride Street
London EC4A 4DL
Great Britain

International metals dealer.

Lonconex Ltd.
29 Mincing Lane
London EC4A 4DL
Great Britain

International metals dealer.

Manfra Tordella & Brookes Inc.
151 World Trade Centre
Concourse
New York, N.Y. 10048
United States

Regional precious metals and numismatics dealer.

McLeod Young Weir Limited
P.O. Box 433
Commercial Union Tower
Toronto Dominion Centre
Toronto, Ontario M5K 1M2
Canada

Major brokerage firm; sells bullion certificates, mining shares, futures and options.

Merrill Lynch Pierce Fenner & Smith Inc.
1 Liberty Plaza
165 Broadway
New York, N.Y. 10006
United States

Major international brokerage firm; sells bullion, futures, options and mining shares.

Midland Doherty Limited
P.O. Box 25
Commercial Union Tower
Toronto Dominion Centre
Toronto, Ontario M5K 1B5
Canada

Major brokerage firm; sells bullion certificates, mining shares, options and futures.

Mocatta & Goldsmid Ltd.
16 Finsbury Circus
London EC2M 7DA Great Britain

Major international bullion dealer; member London gold market.

Mocatta Metals Corporation
25 Broad Street
New York, N.Y. 10004
United States

Major international metals dealer; distributor of major bullion coins; issuer of precious metals options.

Monex International Ltd.
4910 Birch Street
Newport Beach, Cal. 92260
United States

Regional precious metals and numismatics retailer.

Moscow Narodny Bank Limited
24-32 King William Street
London EC4P 4JS
Great Britain

Major intermediary for USSR gold sales to London market.

National Bank of Canada
500 Place d'Armes
Montreal, P.Q. H2Y 2W3
Canada

Issues gold, silver certificates.

Perpetual Storage Inc.
3322 South 300 East
Salt Lake City, Utah
United States

Private vault facility.

Republic National Bank of New York
452 Fifth Avenue
New York, N.Y. 10018
United States

Major international bullion dealer; official distributor for major bullion coins.

Rhode Island Hospital Trust
1 Hospital Trust Plaza
Providence, Rhode Island 02903
United States

Issues gold, silver certificates.

Richardson Greenshields of Canada Limited
One Lombard Place
Winnipeg, Manitoba R3B 0Y2
Canada

Major brokerage firm; sells bullion certificates, mining shares, futures and options.

N.M. Rothschild & Sons Limited
New Court
St. Swithin's Lane
London EC4 Great Britain

Major international bullion dealer; member London gold market.

Samuel Montagu & Co. Ltd.
114 Old Broad Street
London EC2P 2HY
Great Britain

Major international bullion dealer; member London gold market.

The Security Center
147 Carondelet Street
New Orleans, LA 70130
United States

Private vault facility.

Sharps Pixley Ltd.
34 Lime Street
London EC3
Great Britain

Major international bullion dealer; member London gold market.

Shearson/American Express Inc.
2 World Trade Center
New York, N.Y. 10048
United States

Major financial services retailer; sells precious metals accumulation accounts.

Stategic Metals Corporation
500 Chesham House
150 Regent Street
London W1R 5FA
Great Britain

Strategic metals dealers, offering managed accounts.

Strategic Metals Trust
Box 157
CH-1211 Geneva
Switzerland

Strategic metals unit trust.

Summa International Bank Ltd.
Regional Headquarters
12th Floor, BPI Building
Ayala Avenue
Makati, Metro Manila
Philippines

Gold passbook savings accounts.

Swiss Bank Corporation
Aeschenvorstadt 1
CH 4002 Basle
Switzerland

Major international bullion and coin dealer; member of Zurich gold pool.

Swiss Security Systems
4475 South West 8th Street
Miami, Fla. 33134 United States

Private vault facility.

Toronto Dominion Bank
P.O. Box 1
Toronto Dominion Centre
Toronto, Ontario M5K 1A2
Canada

Issues gold, silver certificates.

The Tyndall Group
Box 1256
Hamilton 5
Bermuda

Investment management firm; offers Gold Assurance Fund.

Union Bank of Switzerland
45 Bahnhofstrasse
CH 8021 Zurich
Switzerland

Major international bullion and coin dealer; member of Zurich gold pool.

Valeurs White Weld S.A.
1 Quai du Mont-Blanc
Case Postale 813
CH 1211 Geneva
Switzerland

Major gold, silver options issuer and dealer.

West Coast Bank
Gold and Silver Loan Division
16311 Ventura Boulevard
Encino, Cal. 91436
United States

Specializes in gold, silver margin financing.

Rudolf Wolff & Co. Ltd.
2nd Floor, E Section
Plantation House
10-15 Mincing Lane
London EC3M 3DB
England

Major metals dealer; sells options and futures.

Wood Gundy Ltd.
P.O. Box 274
Royal Trust Tower
Toronto Dominion Centre
Toronto, Ontario M5K 1M7
Canada

Major brokerage firm; sells bullion certificates, mining shares and options.

Wozchod Handelsbank AG
Schützengasse 1
CH 8023 Zurich
Switzerland

Major intermediary in Soviet gold sales to Zurich.

Other Sources of Information

1. ASSOCIATIONS, AGENCIES AND PUBLICATIONS

American Bureau of Metals Statistics Inc.
420 Lexington Avenue
Room 420
New York, N.Y. 10170
U.S.A.

American Metal Market
7 East 12th Street
New York, N.Y. 10003
U.S.A.

British Non-Ferrous Metals Federation
6 Bathurst Street
London W2 2SD
Great Britain

Bureau of the Mint
501 13th Street N.W.
Washington, D.C. 20220
U.S.A.

Chamber of Mines of South Africa
P.O. Box 809
Johannesburg 2000
South Africa

Commodities Research Unit
26 Red Lion Square
London WC1 4RL
Great Britain

Commodity Research Bureau Inc.
1 Liberty Plaza
New York, N.Y.
U.S.A.

Consolidated Gold Fields, PLC
49 Moorgate
London EC2R 6BQ
Great Britain

Department of Energy, Mines & Resources
Government of Canada
580 Booth Street
6th Floor
Ottawa, Ontario K1A 0E4
Canada

Department of the Treasury
15th & Pennsylvania Avenue N.W.
Washington, D.C. 20220
U.S.A.

General Services Administration
General Services Building
18th and F Street N.W.
Washington, D.C. 20405
U.S.A.

The Gold Information Center
645 Fifth Avenue
Olympic Tower
New York, N.Y. 10022
U.S.A.

The Gold Institute
1001 Connecticut Avenue N.W.
Washington, D.C. 20036
U.S.A.

Hambros Bank
Foreign Exchange and Bullion
Dealers Directory
41 Bishopsgate
London EC2P 2AA
Great Britain

International Gold Corporation
645 Fifth Avenue
Olympic Tower
New York, N.Y. 10022

International Monetary Fund
700 19th Street
Washington, D.C. 20431
U.S.A.

International Precious Metals Institute
2254 Barrington Road
Bethlehem, PA 18018
U.S.A.

Minor Metals Traders Association
69 Cannon Street
London EC4N 5AB
Great Britain

Metals Bulletin Ltd.
45 - 46 Lower March
London SE1 7RG
Great Britain

Metals Week
McGraw Hill Inc.
1221 Avenue of the Americas
New York, N.Y. 10020
U.S.A.

The Northern Miner
7 Labatt Avenue
Toronto, Ontario M5A 3P2
Canada

Ontario Ministry of Natural Resources
Public Service Centre
Room 1640
Whitney Block, Queen's Park
Toronto, Ontario M7A 1W3
Canada

Royal (British) Mint
Pontyclun
Mid Glamorgan CF7 8YT
Wales
Great Britain

Royal Canadian Mint
355 River Road
Vanier, Ontario K1A 0G8
Canada

The Silver Institute
1011 Connecticut Avenue N.W.
Suite 1140

Washington, D.C. 20036
U.S.A.

World Bureau of Metals Statistics
41 Doughty Street
London WC1N 2LF
Great Britain

2. EXCHANGES

Chicago Board of Trade
141 West Jackson Boulevard
Chicago, Illinois 60604
U.S.A.

Chicago Mercantile Exchange
International Monetary Market
444 West Jackson Boulevard
Chicago, Illinois 60606
U.S.A.

Commodity Exchange, Inc.
4 World Trade Center
New York, N.Y. 10048
U.S.A.

European Options Exchange
Dam 21
1021 JS Amsterdam
Netherlands

Hong Kong Commodity Exchange
2nd Floor, Hutchinson House
Harcourt Road
Hong Kong

London Gold Futures Market
Ground Floor, Plantation House
Fenchurch Street
London EC3M 3DX
Great Britain

London Metals Exchange
Plantation House
Fenchurch Street
London EC3M 3AP
England

Mid-American Commodity Exchange
174 West Jackson Boulevard
Chicago, Illinois 60604
U.S.A.

The Montreal Exchange
800 Victoria Square
3rd Floor
Montreal, P.Q. H4X 1E9
Canada

The New York Mercantile Exchange
4 World Trade Center
New York, N.Y. 10048
U.S.A.

Singapore Gold Exchange
28th Floor
Clifford Centre
Raffles Place
Singapore 1

Societe de la Bourse de Luxembourg
Boite Postale 165
L-2011 Luxembourg
Luxembourg

Sydney Futures Exchange
13-15 O'Connell Street
Sydney, N.S.W.
Australia 2000

The Toronto Exchange
Exchange Tower
2 First Canadian Place
Toronto, Ontario M5X 1J2
Canada

The Winnipeg Commodity Exchange
678 - 167 Lombard Avenue
Winnipeg, Manitoba R3B 0V7
Canada

Vancouver Stock Exchange
609 Granville Street
P.O. Box 1033
Vancouver, B.C. V7Y 1H1
Canada

Trading Units for Precious and Strategic Metals

NORTH AMERICA'S MOST POPULAR
SMALL BARS AND WAFERS

The following refining firms and banks offer a comprehensive range of small wafers and bars to North American investors. Leading bullion dealers generally buy these products without deducting assay costs.

Product	Gold	Silver	Platinum	Palladium
Credit Suisse/Valcambi S.A.	●	●	●	●
Deutsche Gold-und Silberscheide anstalt ("Degussa")	●	●		
Engelhard Industries	●	●		
Handy & Harman		●		
Johnson Matthey	●	●	●	●
Swiss Bank Corporation /Metaux Precieux S.A.	●	●		
Union Bank of Switzerland /Argor S.A.	●	●		

THE WORLD'S LEADING REFINERS AND ASSAYERS

Gold

AUSTRALIA
Engelhard Industries Limited
Matthey Garrett Pty Limited
The Perth Mint

BELGIUM
Johnson Matthey R. Pauwels S.A.
N.V. Metallurgie Hoboken — Overpelt S.A.
Societe Generale Metallurgique de Hoboken

CANADA
Canadian Copper Refiners Ltd.
Engelhard Industries of Canada
Johnson Matthey Limited
Royal Canadian Mint

CHINA
Refinery of China

FRANCE
Caplain Saint-Andre S.A.
Compagnie des Metaux Precieux
Comptoir Lyon — Alemand Louyot
Laboratories Boudet & Dussaix
Les Anciens Etablissements Leon Martin
Marrett, Bonnin, Lebel & Guieu

GERMANY
Deutsche Gold-und Silber Scheideanstalt
  (Degussa)
W.C. Heraeus GmbH
Norddeutsche Affinerie

GREAT BRITAIN
Bank of England
Engelhard Industries Limited
Johnson Matthey Chemicals Limited
N.M. Rothschild & Sons
The Royal Mint
The Sheffield Smelting Co. Ltd.

ITALY
Metalli Preziosi S.p.A.

JAPAN
Tanaka Kikinzoku Kogyo K.K.

KOREA
Central Bank, D.P.R. of Korea

NETHERLANDS
H. Drijfhout & Zoon's Edelmetaalbedrijven B.V.
Schone Edelmetaal B.V.

PHILIPPINES
Central Bank of the Philippines

SOUTH AFRICA
Rand Refinery Limited

SPAIN
Industrias Reunidas Minero — Metalurgicas
  S.A. (Indumetal)

SWITZERLAND
Argor S.A./Union Bank of Switzerland
Bureau Central Suisse, Controle Metaux
Metaux Precieux S.A./Swiss Bank Corporation
Monnaie Federale Suisse
Valcambi S.A./Credit Suisse
Usine Genevoise de Degrossissage d'Or (Ugdo)

U.S.A.
Asarco Inc.
Engelhard Minerals and Chemicals Corporation
Engelhard Industries
Handy & Harman
Homestake Mining Company
Johnson Matthey Limited
United States Assay Offices & Mint
United States Metals Refining Company

U.S.S.R.
State Refineries

Silver

AUSTRALIA
The Broken Hill Associated Smelters Pty Ltd.
The Electrolytic Refining and Smelting
  Company of Australia

BELGIUM
Metallurgie Hoboken — Overpelt

BURMA
No. 1 Mining Corporation

CANADA
Canadian Copper Refiners Limited
Cominco Limited
Engelhard Industries of Canada Limited
International Nickel Co. (Inco)
Johnson Matthey Limited
Kam-Kotia Mines Limited

FRANCE
Compagnie des Metaux Precieux
Comptoir Lyon-Alemand Louyot

GERMANY, EAST
VEB Masfeld Kombinat Wilhelm Pieck

GERMANY, WEST
Deutsche Gold und Silber Scheideanstalt
Norddeutsche Affinerie

GREAT BRITAIN
Britannia Lead Co. Ltd.
Engelhard Industries Limited
Johnson Matthey Chemicals Ltd.
Royal Mint
Sheffield Smelting Co. Ltd.

ITALY
Metalli Preziosi S.p.A.

JAPAN
Chugai Mining Kogyo Co. Ltd.
Dowa Mining Co. Ltd.
Furukawa Metals Co., Ltd.
Mitsubishi Metal Corporation
Mitsui Mining & Smelting Co., Ltd.
Nippon Mining Co., Ltd.
Sumitomo Metal Mining Co., Ltd.
Toho Zinc Co., Ltd.

MEXICO
Compania Real del Monte y Pachuca
Industrial Minera Mexico

Met-Mex Penoles, S.A.
Sociedad Afinadora de Medales

NETHERLANDS
N.V. Metallurgie Hoboken — Overpelt S.A.

PERU
Cerro de Pasco
Empresa Minera del Centro del Peru

POLAND
Zaklady Metalurgiczne "Trzebinia"

SOUTH AFRICA
Rand Refinery Limited

SPAIN
Industrial Reunidas Minero — Metalurgicas S.A.

SWEDEN
Boliden Aktiebolag

SWITZERLAND
Argor S.A./Union Bank of Switzerland
Metaux Precieux S.A./Swiss Bank Corporation
Valcambi S.A./Credit Suisse

U.S.A.
Ag-Met Refining Corporation
Amax Inc.
American Chemical and Refining Company, Inc.
The Anaconda Company
Asarco Incorporated
The Bunker Hill Company
Engelhard Industries
Engelhard Minerals & Chemicals Corporation
Fisarco Inc.
Handy & Harman
Irvingston Smelting & Refining
Johnson Matthey Limited
Kennecott Corporation
Selby Gold & Silver Refinery
Spirel Metal Co. Inc.
U.S. Assay Office
United States Metals Refining Co.
U.S. Mint

U.S. Smelting, Refining, Mining

U.S.S.R.
Alaverdi, Transcaucasus
Almalyk, Kazakhstan
Norilsk, Eastern Siberia
Novosibirsk, Western Siberia
YUGOSLAVIA
Rudarsko Metalursko-Hemijski Olova: Cinka "Trepca"

Platinum and Palladium

CANADA
Engelhard Industries of Canada Inc.
Falconbridge Nickel Mines
International Nickel Company (Inco)
Johnson Matthey Limited

FRANCE
Comptoir Lyon-Allemand Louyot

GERMANY
Deutsche Gold-und Silberscheideanstalt (Degussa)

GREAT BRITAIN
Engelhard Industries Limited
Johnson Matthey Chemicals Ltd.
Impala Platinum Mines

JAPAN
Tanaka Kikinzoku Kogyo

SOUTH AFRICA
Johnson Matthey
Impala Platinum Mines

SWITZERLAND
Argor S.A./Union Bank of Switzerland
Metaux Precieux S.A./Swiss Bank Corporation
Valcambi S.A./Credit Suisse

U.S.A.
Engelhard Industries Limited
Handy and Harman
International Laboratories
Johnson Matthey Limited
Ledoux & Co.

Lucius Pitkin Inc.
Pennsylvania Massey Bishop Inc.
PGP Industries Inc.
U.S.S.R.
Amtorg

TRADING UNITS FOR STRATEGIC METALS

Metal	Purity Grades	Comments	Trading Units	Most Negotiable
Antimony	99.90-99.85%	maximum arsenic content 0.2%	Ingots, 20-50 lbs.	99.65% material with less than 0.15% arsenic
Beryllium	usually 5% in copper alloy	pure beryllium too toxic to be suitable	various, depends on alloy	alloy of 5% beryllium 95% copper
Cadmium	99.95-99.99%		2 inch balls 10 inch sticks ingots	99.99% sticks to be used for chemical purposes
Chromium	differs widely according to purpose	mostly used in ore form, not purely	ferro-chrome lumps in drums, chromite in bulk	bulk ferro-chrome of about 55% purity
Cobalt	99.60% minimum		broken cathodes, granules	broken cathodes, min. 99.6% pure
Columbium (or Niobium)	usually 60-70% in ferro-alloy	pure columbium not often produced	ferro-columbium in lumps, packed in drums	bulk ferro-columbium of about 65% purity

Metal	Purity Grades	Comments	Trading Units	Most Negotiable
Gallium	99.90-99.9999%		small ingots	high purity ingots
Germanium	99.999%	grades often determined by electrical resistance, not by purity	ingots, single crystals	ingots of 30 ohm cms or 50 ohm cms (el. res.)
Indium	99.97-99.99%		bars, ingots	ingots or bars of 99.99% purity
Lithium	depends on compound	pure lithium rarely produced	in carbonate, hydroxide, or fluoride compounds	lithium - carbonate
Magnesium	99.8-99.9%		ingots of 5-20 kg	99.9% pure ingots
Manganese	99.8-99.9%	used purely and as alloys	electrolytic lumps, bulk ferro-manganese 80% bulk silicon-manganese 80%	99.9% pure lumps
Mercury	99.99%		flasks, 76 lbs/ bottle	99.99% pure

Trading Units for Strategic Metals

Metal	Purity Grades	Comments	Trading Units	Most Negotiable
Molybdenum	60-80% in ferro-alloy	hardly traded in metallic form. Mostly sold as oxide or alloy	ingots, sheets, wire	bulk ferro-molybdenum of 60-80% purity
Selenium	99.5-99.8%	powder should be "minus 200 mesh" or finer	powder, granules	fine powder of 99.8% purity
Silicon	98.5% or more	maximum iron content 0.5%, calcium content 0.3%	irregular lumps	98.5% pure with less than 0.5% iron, less than 0.3% calcium
Tantalum	usually 30%, 60% in tantalite, or 99.9% with pure metal	mostly traded in ore form ("tantalite")	bulk tantalite, powder, sheets or bars	powder of 99.9% purity for use in electronics
Tellurium	99.7-99.9%		small ingots, slabs	ingots of 99.9% purity

Metal	Purity Grades	Comments	Trading Units	Most Negotiable
Titanium	99.6%		sponge, slabs, bars, sheets, briquettes, wire	sponge or briquettes 99.6% pure
Tungsten	depends on concentrate	pure tungsten not frequently traded	metric ton units in concentrates	"wolfram-ore" containing about 68% tungsten oxide
Vanadium	usually 50% or 80% in ferro-vanadium	pure vanadium seldom traded	vanadium pentoxide ferro-vanadium	bulk ferro-vanadium
Zirconium	depends on use	used in a variety of alloys	sponge, ingots	sponge or ingots

Weights and Weight Conversions

Troy Ounces and Grams

One troy ounce is equivalent to 31.1034807 grams but, in practice, this conversion factor is usually rounded off to
31.1035

1 troy ounce	=	31.1035 grams
2 troy ounces	=	62.2070 grams
5 troy ounces	=	155.5175 grams
10 troy ounces	=	311.0350 grams
50 troy ounces	=	1,555.1750 grams
100 troy ounces	=	3,110.3500 grams
400 troy ounces	=	12,441.4000 grams
1,000 troy ounces	=	31,103.5000 grams

Conversely, one gram equals 0.0321507 troy ounces (1 ÷ 31.1035) but, in practice, this conversion factor is usually rounded off to
0.03215

1 gram	=	0.03215 troy ounces
5 grams	=	0.16075 troy ounces
10 grams	=	0.32150 troy ounces
20 grams	=	0.6430 troy ounces
50 grams	=	1.60750 troy ounces
100 grams	=	3.2150 troy ounces
500 grams (½ kilogram)	=	16.0750 troy ounces
1,000 grams (1 kilogram)	=	32.1500 troy ounces

One troy ounce equals 1.09711 Avoir Dupois ounces.

Conversion Table for various weights of precious and strategic metals

Vertical readings will tell you what each unit contains

Each unit contains →	Grain	Gram	Pennyweight	Tola	Ounce	Troy Ounce	Tael	Pound	Kilogram	Short Ton	Metric Ton	Long Ton
Grain		15.430	24	180	437.52	480	577.63	7,014	15,432	n/p	n/p	n/p
Gram	0.0647		1.5552	11.664	28.351	31.1035	37.430	454.53	999.98	n/p	n/p	n/p
Pennyweight	0.0416	0.6430		7.5	18.23	20	24.068	292.27	643	n/p	n/p	n/p
Tola	0.00555	0.08573	0.1333		2.4307	2.6667	3.2091	38.970	85.734	77,798	85,734	87,129
Ounce	0.00228	0.03527	0.0549	0.4114		1.0971	1.3203	16.000	35.272	32,007	35,272	35,846
Troy Ounce	0.00208	0.03215	0.0500	0.3750	0.9115		1.2024	14.6136	32.150	29,174	32,150	32,673
Tael	0.00173	0.02672	0.0415	0.3116	0.7574	0.8310		12.144	26.717	24,244	26,717	27,151
Pound	0.00014	0.00220	0.0034	0.0257	0.0623	0.0684	0.0823		2.2000	1,995.5	2,199	2,234.8
Kilogram	n/p	0.00100	0.0016	0.0117	0.0283	0.0311	0.0374	0.4545		907.31	999.87	1,016.1
Short Ton	n/p	n/p	n/p	n/p	n/p	n/p	n/p	0.0005	0.0012		1.1022	1.1200
Metric Ton	n/p	n/p	n/p	n/p	n/p	n/p	n/p	0.0004	0.0010	0.9073		1.016
Long Ton	n/p	n/p	n/p	n/p	n/p	n/p	n/p	0.0004	0.0010	0.8928	0.9842	

Read horizontally to find out how many units are contained in each foreign unit.

n/p = conversions are not practical

Precious metals weights applied to finenesses

Gross Weight	Fine Content in Troy Ounces		
	0.9950	0.9990	0.9999
1 kilogram	31.990	32.119	32.148
½ kilogram	15.995	16.059	16.074
200 grams	6.398	6.424	6.430
100 grams	3.199	3.212	3.215
50 grams	1.600	1.607	1.608
20 grams	0.640	0.643	0.643
10 grams	0.321	0.322	0.322
5 grams	0.161	0.161	0.161
100 troy ounces	99.500	99.900	99.990
50 troy ounces	49.750	49.950	49.995
20 troy ounces	19.900	19.980	19.998
10 troy ounces	9.950	9.990	9.999
5 troy ounces	4.975	4.995	5.000
2 troy ounces	1.990	1.998	2.000
1 troy ounce	0.995	0.999	1.000
10 tolas	3.731	3.746	3.750
5 tolas	1.866	1.873	1.875
1 tola	0.373	0.375	0.375
10 taels	11.974	12.022	12.032
5 taels	5.987	6.011	6.017
1 tael	1.197	1.202	1.203

Fineness Expressed in Karats

The jewelry industry expresses the fineness of a product in karats. The conversion factor is always 24, in other words, a product of the highest possible purity is 24 karats fine.

24 karats = 1000 (or 999.9) parts fine
22 karats = 916.7 parts fine
18 karats = 750.0 parts fine
14 karats = 583.3 parts fine
10 karats = 416.7 parts fine
 9 karats = 375.0 parts fine
 8 karats = 333.3 parts fine

The fineness of a metal is usually expressed in the ratio between pure parts out of a total of 1,000 parts contained in an alloy.

Metals Talk — A Guide to the Terminology

Acid Test
The fineness of certain metals can be determined by exposure to various acids: higher quality gold, for instance, can usually be tested by exposure to aqua regia, a combination of nitric and hydrochloric acids.

Actuals
Used to describe a physical metal or commodity, as opposed to a futures contract.

ADR
Abbreviation for American Depositary Receipt. Issued by a U.S. bank, an ADR represents one or more units of a foreign security and is generally issued to simplify the physical handling and the legal technicalities governing foreign securities issues. Australian and South African mining shares traded in the U.S., for instance, are generally in the form of ADR's.

Alloy
A mixture of metals. Platinum or palladium, for instance, are used predominantly as alloys. Gold is often alloyed with copper, nickel, silver or zinc to improve its hardness or to change its colour.

Alluvial Gold
Placer gold, formed by the weathering of gold bearing rocks, and carried downstream by rivers.

Arbitrage
To simultaneously buy and sell a commodity or security in different markets. An "arbitrageur" makes money by purchasing gold in Chicago and selling it at the same time in Winnipeg, taking advantage of a price differential. Such differences in the market price are usually so small that arbitrage trading requires substantial amounts to generate a profit.
The term arbitrage is also used for a whole string of complicated trading manoeuvres, exploiting differences in spot prices, futures prices and interest rates.

Assay, Assayer
An assay is a test of a metal in which its purity is

determined; an assayer has the function of conducting this analysis and usually confirms it by marking a precious metal accordingly.

Bid
Opposite of "offer". The firm price quoted by an interested buyer.

Bull and Bear
Someone is called a "Bull" when he expects the market to rise and a "Bear" when he expects the market to fall.

Bullion
Precious metal in negotiable form; in most marketplaces a purity of .995 or finer is required. Bullion is produced in the form of bars, wafers or ingots.

Call option
By paying a premium, the option buyer acquires the right to decide at a later date whether or not to buy a metal or a commodity at a pre-arranged price.

Carat
See "karat".

Carrying Cost
The cost connected with owning a metal. This can include the loss of interest, the warehouse storage, the insurance, as well as estimated changes in weight.

Cash Price
Also called "spot price". The price required for immediate settlement. The term cash or spot is used to differentiate from a futures transaction, where settlement is due at a predetermined time in the future.

Centenario
Mexican gold piece with a face value of 50 pesos. First issued in 1921 for the 100th anniversary of Mexico's independence.

Chervonetz
Russian gold piece with a face value of 10 rubles. Modern chervonetz were first minted in 1923 as legal tender, and were minted in far greater quantity after 1975.

Closing out
Liquidating or offsetting an existing long or short futures position, also known as "exiting".

Coin Gold
An alloy generally used for the production of gold coinage. In order to improve the durability, gold is usually mixed with silver and

copper. Many European coins have a fineness of .75 (18 k), while in the U.S. a fineness of .90 (21.6 k) is used.

Cover

A trader who is short a commodity in the futures market offsets his obligation by buying the same quantity. This action is also called "short covering".

Deferred Contract

A futures contract with a distant delivery date.

Delivery Month

This term is used by futures traders to describe the contract of a specific commodity with the nearest possible delivery date. The "delivery month" is also frequently called "spot month".

Dental Gold

Gold of usually 16 or 20 karats.

Discount

Mostly used in futures trading. A commodity may trade "at a discount" for delivery at a future time. This means that its price is quoted lower than if you were to purchase it today.

Doré Bullion

An impure alloy of a precious metal produced at a mine.

Double Eagle

United States gold coin with a face value of $20. An Eagle was a $10 gold piece, which explains why the $20 coin is called a Double Eagle.

Ducat

Term traced back to the inscription on coins circulated in 12th century Italy, and since used for a variety of European coins.

Ductility

The ability of a metal to be beaten into a thin sheet or to be drawn into fine wire.

EFP

An acronym for "exchange for physicals". An EFP is a trade between two parties in which one party is a buyer of physical metals and the seller of the equivalent in the form of a futures contract, and the other party is the seller of physical metals and the buyer of the same quantity in the form of a futures contract.

Electrolysis

A refining process in which a metal is deposited on a cathode from a solution or a molten mass.

Element

A substance which cannot be split into anything simpler in a chemical process.

Face Value   Also known as "nominal value". The monetary value given to its currency or coin by a government, so that it may be used as legal tender.

Fineness   Refers to fine metal content in 1000 parts of an alloy. A silver bar of a fineness of .999 (pronounced "three nines") contains 999 parts of pure silver and one part of something else. Fineness is also expressed in karats, especially in the jewelry trade. The fineness is usually stamped along with the assayer hallmark.

Fine Weight   The weight of pure precious metals contained in a bar or coin.

Fixings   In the London market, the price at which dealers transact gold and silver with each other is fixed. In the case of gold, the five market participants get together twice every day for a "fixing", while a similar procedure takes place once a day for silver. The London fixings are quoted and observed around the world as important indicators of the market trend.

Flat   "Being flat", or "flat position" refer to a trader's net position in a commodity, meaning that his books show no holdings or liabilities. The term is used mostly in futures markets.

Fool's Gold   Iron pyrite is often mistaken for gold. The qualities of gold and iron pyrite are quite different: real gold is soft and malleable, the false version is hard and brittle.

"Four Nines"   The finest gold bullion or gold coinage available is gold with a fineness of .9999, or as the experts call it, "four nines fine gold".

Futures Contract   A contract between a buyer and seller of a commodity or security, setting its price and date of delivery. If you purchase gold in the futures market, you undertake to pay a certain amount of money for it at an agreed date in the future.

Gilding   A gilded surface is a surface coated with a thin layer of gold. Today most surfaces are electro-plated.

Gold Leaf

Gold hammered to extremely thin consistency. Gold leaf is used for a variety of purposes, most of which are decorative. Picture frames and book edges are the most common applications of gold leaf. The art of producing gold leaf dates back to ancient Egyptian times.

Gold Parity

Value of a currency unit expressed in milligrams of fine gold.

Gold Standard

A monetary system based on convertibility into gold, e.g. a nation issues paper currency and fully backs it with gold. The two "moneys", gold and paper, are then freely interchangeable in terms of each other.

Good Delivery

"Good delivery bars" are bars meeting the specifications established by an exchange or by the dealer community.

Grain

Originally the equivalent of one grain of wheat or barley, the grain is the earliest weight unit for gold. The "grain" is actually still used in the Avoirdupois System and in the Troy System. One grain is equivalent to 0.0648 grams, 24 grains make one pennyweight (dwt.) and one pound troy consists of 5,769 grains.

Gram

1/1000 of one kilogram. The continental European gold market is dominated by metric weights (kilograms, grams); the British and North American markets by troy ounces.
1 gram = 15.43 grains = 0.032 ounces troy
1 pound troy = 373.2 grams
32.15 ounces troy = 1,000 grams = 1 kilogram

Hallmark

A mark or a number of marks, which indicate the producer of a bar and its particulars (number, fineness, etc.) Originally used in England as early as 1300 A.D.

Hedge

A transaction by a consumer or producer of a metal designed to protect him against price fluctuations. A consumer of platinum, for instance, may "hedge" against a possible price increase by buying enough metal to cover his needs in the form of a futures contract. Futures markets were originally for hedgers, as opposed to speculators, by whom the market is now used predominantly.

Ingot
A term originally used for standard delivery gold or silver bars. Today, "ingot" is used for any negotiable bar.

Karat
The name karat originates from Greece and comes from the word originally used for the carob seed, which was used as a uniform weight. With precious metals, karat is a measure of purity. Totally pure gold, for instance, is 24 karat.

Karat Gold
Gold of not less than 10 k. fineness.

Krugerrand
The world's best selling bullion type coin. Minted since 1967, Krugerrand coins are now available in sizes of one troy ounce, as well as 1/2, 1/4 and 1/10 of a troy ounce.

Leverage
The ability to make an investment without paying its full nominal value. If, for instance, you purchase 100 ounces of platinum in the futures market, you need only deposit around 10% of total value.

Limit Down
Limit Up
Trading in futures contracts is governed by certain limits set by the exchange. Silver traded on New York's COMEX, for instance, cannot move more than $0.50 per ounce in either direction in one trading day.

Liquid Gold
Usually used for surface decoration such as glass. A solution of gold (normally 12%) and a mixture of chemicals is painted on and then heated to a temperature of 540 degrees Celsius. The resulting coating is only 0.1 microns thick (0.00004 inches).

Locals
Exchange members who basically trade for their own account. Some locals also execute

orders for other members by acting as "floor brokers".

Loco
The place at which a certain metal lies, or will be delivered. You may hold antimony "loco London", or may want to buy tungsten for delivery "loco New York".

London Delivery Bar
Gold Trading unit used in the London Gold Market.
A London Delivery Bar (also "standard bar") is a bar weighing approximately 400 ounces troy, having a minimum fineness of .955 and carrying the markings of a melter or assayer listed on the "good delivery list" 'those refiners and assayers whose markings are internationally recognized'.

Long
"Going long" or "long position" describes the purchase of a commodity or security. Usually, the term is used for futures market transactions. To go long palladium, for instance, means to purchase palladium in the hope of a price increase.

Lot
The standard contract unit on a regulated market. On New York's COMEX, for instance, a "silver lot" is equivalent to 5,000 troy ounces. In the stock markets, a "lot" usually means 100 or 1,000 shares. A purchase of 65 shares, on the other hand, would be an "odd lot".

Malleability
Some metals can be extended considerably by hammering and do not crack or break in the process. Gold is the most malleable metal. It can be hammered to a thickness of only 0.000005 inches. One ounce of gold, in fact, can be beaten into a sheet covering over 100 square feet.

Maple Leaf
One of the world's best selling gold coins. Maple Leafs were first minted in 1979 as one troy ounce coins with a face value of $50. Today, there are also fractional "Maple Leafs" in sizes of 1/2 ounce and 1/10 ounce.

Margin
A security deposit left with a broker to protect

him from possible losses incurred as a result of a long or short position you may hold in the futures market.

Margin Call
A commodity broker's request to a client for additional funds.

Marks
The stamp or stamps attached to bullion ingots by producers or by assayers. Producers usually mark a bar with its fineness, its weight, a bar number, their hallmark and a definition of the metal, such as "platinum" or "fine silver". An assayer, testing only the fineness of a bar, will either stamp his logo under the fineness stamp of the original producer, or will stamp on the fineness again.

Melter
Also referred to as "the refiner". The melter buys raw material from mines, refines it to an acceptable purity and then marks the finished bullion accordingly.

Napoleon
Originally the name of a French gold piece bearing the image of Napoleon I or III, this term has become generic for all subsequent French gold coins.

Offer
Opposit of "bid". The firm price quoted by an interested seller.

Open Interest
The total number of commodities contracts on an exchange which have not been offset by the opposite transaction or by physical delivery.

Open Outcry
Trading conducted by calling out bids and offers across a ring or pit and having them accepted in the same manner.

Option
The right to purchase or sell a commodity or security at a specified time in the future. You usually acquire this right at a certain price or are paid a certain price for it.

Pennyweight
Used in the United States as a unit of weight for gold. In Troy weight, twenty pennyweights equal one ounce. The term pennyweight dates back to England where, in the 17th century, it

referred to the weight of one silver penny or 1/24th of a Tower Pound.

Physical

Metals bought in actual bullion form are called physical metals, or actual metals, in contrast to paper transactions, such as futures or options.

Placer

Used for all mineral deposits which are loose and not "in place". Platinum found in rivers, for instance, is placer platinum.

Premium

In futures, the amount by which the price of one commodity contract sells above another. For instance, palladium for delivery in May may trade at a premium over palladium due in February.

In options, the amount paid for the "right" to buy or sell a commodity at a prearranged price. Premium is also used to express, as a percentage, the amount by which the market value of a precious metal wafer or coin exceeds the value of the actual metal it contains. If the gold contained in a bullion coin is worth $400 but the coin sells for $416, then that coin is trading at a premium of four percent.

Primary metal

Refers to unworked metals. The raw product from a mining operation after the ore has been separated.

Put option

In exchange for a premium, the buyer of a put option acquires the right to decide at a later date whether or not to sell a commodity at a pre-arranged price.

Rollover

The process of moving forward a futures position that nears the delivery date. When the rollover is executed, the old position is liquidated, and a new position in a further out delivery month is simultaneously established.

Scrap

Any kind of precious metal not in the form customarily traded. The term scrap is primarily used for industrial waste material and melted jewelry.

Short

A trader goes short when he sells a commodity

for future delivery that he does not own. He does so in the hope of repurchasing the same commodity at a lower price some time before the delivery date.

Smelting A process of extracting crude metal from its ore using heat. The smelting process takes place as a preliminary step to refining.

Sovereign Traditional English gold coin which always bears the likeness of the reigning monarch, or sovereign.

Solid Gold Defined by the U.S. Federal Trade Commission as "any article that does not have a hollow center and has a fineness of ten karats or higher." Investors, beware!

Sponge A lumpy form of metal with a sponge-like appearance produced by casting molten metal into water. Platinum and palladium are often traded in the form of sponge for industrial purposes.

Spot Price See "cash price".

Spread The difference between the buying and selling price. If platinum is bought at $400 and sold at $402, the spread is two dollars. The term is used differently by futures contracts traders. See "straddle".

Standard Bar See "London delivery bar."

Sterling Silver of a fixed standard of purity, usually 925 parts of silver with 75 parts of copper. Sterling is used primarily for ornamental purposes, e.g. sterling tableware, etc.

Stoploss The placement of a buying order or selling order against an existing futures contract. A stoploss is designed to limit potential loss. After buying a palladium contract of 100 ounces at $120, you may decide the maximum loss you want to incur is $1,000. You therefore place a stoploss at $110, meaning your broker will automatically sell your contract if that price is reached.

Straddle

Used mostly in futures trading. A trader may find it advantageous to sell July gold and purchase simultaneously an identical amount of November gold, exploiting the price differential. Traders call this maneuver a straddle or a spread.

Striking Price

Used mainly in options trading. The striking price is the price at which a commodity will change hands if an option is exercised.

Tael

Chinese gold weight; 1 tael = 1.2034 ounces troy of .9999 fineness. Taels were almost totally restricted to Far Eastern markets until the end of the Vietnam War when thousands of refugees brought them to North America and Europe.

Tola

Old gold weight. One tola equals 3.750 ounces or 111 grams, of .9999 fineness.

Troy Ounce

A unit of the troy system of weight used for the measurement of precious metals. Gold, silver, platinum and palladium are expressed in troy ounces.
1 troy ounce = 1.09711 ounces or 31.103 grams.

Wafer

Term used for bullion manufactured in thin form, e.g., one ounce silver wafer.

Warehouse Receipt

Issued when a futures contract expires and delivery is taken. Evidence that a commodity is held in physical form in an exchange approved warehouse.

Year Mark

A letter symbol stamped onto gold by British goldsmiths. Each year has a different mark.

Recommended Reading

The following is a selection of what I believe to be the most useful and important books written on strategic and precious metals.

1. **INVESTMENT IN PRECIOUS AND STRATEGIC METALS**

Bandulet, Bruno, "Gold Guide", Fortuna-Finanzverlag, Switzerland, 1984.

Browne, Harry, and **Coxon**, Terry, "Inflation — Proofing Your Investments", Morrow & Co., New York, 1981.

Casey, Douglas, "Crisis Investing", '76 Press, Seal Beach, Ca., 1979.

Cavelti, Peter C., "How To Invest In Gold", Maximus Press/McClelland & Stewart, Toronto, and Follett Publishing, Chicago, 1979, 1981.

Chakrapani, Chuck, "Silver: A Guide to Investment Profits", Standard Research Systems Inc., Toronto, 1984.

Chamberlain, Geoffrey, "Trading In Options", Woodhead-Faulkner, Cambridge, England, 1981, 1982.

Day, Adrian, "Investing Without Borders", Alexandria House Books, Alexandria, Va., 1982.

Edwards, John, and **Reidy**, Brian, and **de Keyser**, Ethel, "Guide To World Commodity Markets", Kogan Page, London and Nichols Publishing, New York, 1979.

Edwards, John, and **Robbins**, Peter, "Guide to Non-Ferrous Metals and Their Markets", Kogan Page, London and Nichols Publishing, New York, 1979, 1980.

Lamb, Norman, "Small Fortunes in Penny Gold Stocks", The Penny Mining Prospector, Silver Springs, Md., 1982.

Levine, Ely, "The Golden Key: A Complete Guide to Gold Coin Investment", Purnell & Sons Ltd., Bryanston, South Africa, 1980.

McLendon, Gordon, "Get Really Rich In The Coming Super Metals Boom", Pocket Books, Simon & Schuster, New York, 1980, 1981.

Robbins, Peter, and **Lee**, Douglas, "Guide to Precious Metals", Kogan Page, London and Nichols Publishing, New York, 1971.

Sarnoff, Paul, "Trading In Gold", Woodhead-Faulkner, Cambridge, England, 1980.

Sherman, Eugene J., "Gold Investment Papers", International Gold Corporation Ltd., New York, 1981, 1982.

Smith, Jerome F. and **Kelly-Smith**, Barbara, "Silver Profits in the Eighties", Books in Focus, New York, 1982.

Swiss Bank Corporation, "Gold", Zurich, 1980.

The Montreal Exchange and **The Vancouver Stock Exchange**, "Understanding Gold Options", Montreal, 1982.

Valeurs White Weld S.A., "Options on Gold Bullion", Geneva, 1977.

2. PRECIOUS AND STRATEGIC METALS IN MONETARY HISTORY

Bresciani-Turroni, Costantino, "The Economics of Inflation: A Study of Currency Depreciation in Post-War Germany", A.M. Kelly, New York, 1968.

Busschau, W.J., "Gold and International Liquidity", South African Institute of International Affairs, Johannesburg, 1961.

Cassell, Francis, "Gold or Credit? The Economics and Politics of International Money", Praeger, New York, 1965.

Einzig, Paul, "The Destiny of Gold", Macmillan, London, 1972.

Gold, Joseph, "Floating Currencies, Gold and SDR's", International Monetary Fund Pamphlets No. 19 and No. 22, Washington, D.C., 1976, 1977.

Harrod, Sir Roy F., "Reforming The World's Money", Macmillan, London, 1965.

Hayek, Friedrich A., "Choice in Currency: A Way to Stop Inflation", Transatlantic, Levittown, N.Y., 1977.

Hazlitt, Henry, "What You Should Know About Inflation", Van Nostrand Reinhold, New York, 1965.

Jastram, Roy W., "Silver, The Restless Metal", John H. Wiley & Sons, New York, 1981.

Jastram, Roy W., "The Golden Constant: The English and American Experience, 1560 - 1976", John H. Wiley & Sons, New York, 1977.

Meiselman, David I., and **Laffer**, Arthur B., "The Phenomenon of Worldwide Inflation", American Enterprise Institute, Washington, 1975.

Rees-Mogg, William, "The Reigning Error: The Crisis of World Inflation", Hamilton, London, 1974.

Triffin, Robert, "Gold and The Dollar Crisis: The Future of Convertibility", Yale University Press, New Haven, Conn., 1961.

Von Mises, Ludwig, "The Theory of Money and Credit", Foundation for Economic Growth, New York, 1971.

3. SPECIALTY BOOKS ON PRECIOUS AND STRATEGIC METALS

Consolidated Gold Fields, "Gold", London, published annually.

Cousins, C.A., "Platinum Group Metals", Department of Mines Geological Survey, Handbook 7, Mineral Resources of South Africa, 1976.

Derry, Duncan R., "A Concise World Atlas of Geology and Mineral Deposits", Mining Journal Books, London, and Halsted Press/John Wiley & Sons, New York, 1980

Gibson-Jarvie, Robert, "The London Metal Exchange", Woodhead-Faulkner, Cambridge, England, 1976, 1983.

Granger, C.W.J., "Trading In Commodities", Woodhead-Faulkner, Cambridge, England, 1974, 1983.

Green, Timothy, "The New World of Gold", Walker & Company, New York, 1982.

Kettell, Brian, "Gold", Ballinger Publishing Co., Cambridge, Mass., 1982.

Mohide, Thomas Patrick, "Gold", Ontario Mineral Policy Background Paper No. 12, Ministry of Natural Resources, Ontario, 1981.

Mohide, Thomas Patrick, "Platinum Group Metals — Ontario and the World", Ontario Mineral Policy Background Paper No. 7, Ministry of Natural Resources, Ontario 1979.

Page, N.J., and **Clark**, A.L., and **Desborough**, G.A., and **Parker**, R.L., "Platinum Group Metals", United States Mineral Resources, U.S. Geological Survey Professional Paper No. 820, 1973.

Samuel Montagu & Co., "Annual Bullion Review", London, published annually.

Sarnoff, Paul, "Silver Bulls", Arlington House Publishers, Westport, Conn., 1980.

Sutton, Anthony C., "Investing in Platinum Metals", Adam Smith Publishing, U.S., 1982.

Szuprowicz, Bohdan, "How to Invest In Strategic Metals", St. Martin's Press, New York, 1982.

Taylor-Radford, R.S., "The Metals Investment Handbook", The Common Sense Press Inc., Costa Mesa, Cal., 1983.

"The London Gold Market", by its members, London, 1980.

U.S. Bureau of Mines, Department of The Interior, "Platinum Group Metals", Washington, D.C., 1980.

Acknowledgements

Before I express my thanks to those who have helped me in getting this book researched, written and published, I should like to note that my studies would not have been nearly as productive without the efforts of hundreds of experts whose comments and materials I have followed over the years. My first acknowledgement of appreciation should therefore go to them.

Thanks also go to my friend Max Layton, who was instrumental in getting me to write my first book, "How To Invest In Gold". His advice and help in reviewing the manuscripts and research for this book were equally invaluable. My loyal assistant, Jan Higgins, spent many hours of overtime in typing and correcting my work, and my colleagues at Guardian Trust Company supported me in letting me use the firm's research facilities. Special thanks go to all of them.

Peter C. Cavelti

Index